IEEE Recommended Practice for Powering and Grounding Sensitive Electronic Equipment

Published by
The Institute of Electrical and Electronics Engineers, Inc.

IEEE Std 1100-1992

IEEE Recommended Practice for Powering and Grounding Sensitive Electronic Equipment

Sponsor

**Power System Engineering Committee
of the
IEEE Industrial Applications Society**

Approved June 18, 1992

IEEE Standards Board

Abstract: Recommended design, installation, and maintenance practices for electrical power and grounding (including both power-related and signal-related noise control) of sensitive electronic processing equipment used in commercial and industrial applications are presented. The main objective is to provide a consensus of recommended practices in an area where conflicting information and confusion, stemming primarily from different view points of the same problem, have dominated. Practices herein address electronic equipment performance issues while maintaining a safe installation. A brief description is given of the nature of power quality problems, possible solutions, and the resources available for assistance in dealing with problems. Fundamental concepts are reviewed. Instrumentation and procedures for conducting a survey of the power distribution system are described. Site surveys and site power analysis are considered. Case histories are given to illustrate typical problems.

Keywords: commercial applications, electrical power, grounding, industrial applications, sensitive equipment

Grateful acknowledgment is made to the following for having granted permission to reprint illustrations in this document as listed below:

Figures 2-1, 2-3, 2-4, 2-5, 4-2(a), and 4-2(b) from *The Dranetz Field Handbook for Power Quality Analysis,* Dranetz Technologies, Inc., Edison, NJ, 1991.

Figures 2-2(a), 2-2(b), and 4-3 from A. McEachern, *Handbook of Power Signatures*, Basic Measuring Instruments, Foster City, CA, 1988.

Table 4-2 from ANSI C84.1-1989, American National Standard for Electric Power Systems and Equipment—Voltage Ratings (60 Hz), copyright 1989 by the American National Standards Institute.

Figure 4-12 from R. H. Golde, ed., *Lightning*, Vol. 2, *Lightning Protection*, London: Academic Press, 1977, p. 802.

Figure 4-15 from R. B. Standler, *Protection of Electronic Circuits for Overvoltages*, John Wiley & Sons, Inc., copyright © 1989.

Figure 5-1 from John Fluke Manufacturing Co., Inc., Everett, WA.

Figure 5-2 from ECOS Electronics Corp., Oak Park, IL.

Figure 5-3 from Oneac Corp., Libertyville, IL.

Figure 5-4 from Monistar Systems, Inc., Richmond, VA.

Figure 5-5 from Superior Electric, Bristol, CT.

Figure 5-6 from Angus Electronics Co., Indianapolis, IN.

Figure 5-7 from Basic Measuring Instruments, Foster City, CA.

Figure 5-8 from Dranetz Technologies, Inc., Edison, NJ.

Figure 5-9 from Telog Instruments, Inc., Henrietta, NY.

Figure 5-10 from Collective Intelligence, Inc., Santa Rosa, CA.

First Printing

The Institute of Electrical and Electronics Engineers, Inc.
345 East 47th Street, New York, NY 10017-2394, USA

Copyright © 1992 by
The Institute of Electrical and Electronics Engineers, Inc.
All rights reserved. Published 1992
Printed in the United States of America

ISBN 1-55937-231-1

No part of this publication may be reproduced in any form, in an electronic retrieval system or otherwise, without the prior written permission of the publisher.

December 31, 1992

IEEE Standards documents are developed within the Technical Committees of the IEEE Societies and the Standards Coordinating Committees of the IEEE Standards Board. Members of the committees serve voluntarily and without compensation. They are not necessarily members of the Institute. The standards developed within IEEE represent a consensus of the broad expertise on the subject within the Institute as well as those activities outside of IEEE which have expressed an interest in participating in the development of the standard.

Use of an IEEE Standard is wholly voluntary. The existence of an IEEE Standard does not imply that there are no other ways to produce, test, measure, purchase, market, or provide other goods and services related to the scope of the IEEE Standard. Furthermore, the viewpoint expressed at the time a standard is approved and issued is subject to change brought about through developments in the state of the art and comments received from users of the standard. Every IEEE Standard is subjected to review at least once every five years for revision or reaffirmation. When a document is more than five years old, and has not been reaffirmed, it is reasonable to conclude that its contents, although still of some value, do not wholly reflect the present state of the art. Users are cautioned to check to determine that they have the latest edition of any IEEE Standard.

Comments for revision of IEEE Standards are welcome from any interested party, regardless of membership affiliation with IEEE. Suggestions for changes in documents should be in the form of a proposed change of text, together with appropriate supporting comments.

Interpretations: Occasionally questions may arise regarding the meaning of portions of standards as they relate to specific applications. When the need for interpretations is brought to the attention of IEEE, the Institute will initiate action to prepare appropriate responses. Since IEEE Standards represent a consensus of all concerned interests, it is important to ensure that any interpretation has also received the concurrence of a balance of interests. For this reason IEEE and the members of its technical committees are not able to provide an instant response to interpretation requests except in those cases where the matter has previously received formal consideration.

Comments on standards and requests for interpretations should be addressed to:

> Secretary, IEEE Standards Board
> 445 Hoes Lane, P.O. Box 1331
> Piscataway, NJ 08855-1331
> USA

IEEE Standards documents are adopted by the Institute of Electrical and Electronics Engineers without regard to whether their adoption may involve patents on articles, materials, or processes. Such adoption does not assume any liability to any patent owner, nor does it assume any obligation whatever to parties adopting the standards documents.

Foreword

(This foreword is not a part of IEEE Std 1100-1992, IEEE Recommended Practice for Powering and Grounding Sensitive Electronic Equipment.)

This Recommended Practice is a publication of the Industry Applications Society of the IEEE and is one of the IEEE Color Book Series, which relates to industrial and commercial power systems. The purpose of the Recommended Practice is to provide consensus for installing and providing power to sensitive electronic equipment, in literally all sectors and power system environments. This has been a growing area of concern as incompatibilities between power system characteristics and equipment tolerances have caused operating problems and loss of productivity in all kinds of power systems.

As load and source compatibility concerns have become more common, the facility engineers and system designers have been in the spotlight to provide solutions. Power and microelectronic equipment designs also have a role in solving the problems. Electronic equipment can be a contributor to, and a victim of, powering and grounding incompatibilities in power systems. A cooperative effort is required among power system designers, equipment manufacturers, and the electric utilities to provide and maintain an acceptable level of load/source compatibility.

To address this multidisciplined area, the Working Group on Powering and Grounding Sensitive Electronic Equipment was formed in 1986 to write a Recommended Practice. The project was sponsored by the IAS Industrial and Commercial Power Systems Department, Power Systems Engineering Committee. This practice is intended to complement other Recommended Practices in the IEEE Color Book Series and has been coordinated with other related codes/standards, as well as recognized testing laboratories.

At the time this Recommended Practice was approved, the IEEE Working Group on Powering and Grounding Sensitive Electronic Equipment had the following membership:

Thomas S. Key, *Chair*
Warren H. Lewis, *Secretary;* **Van Wagner,** *Secretary*
Michael C. Keeling, *Technical Editor*
François D. Martzloff, *Coordinating Editor*

Vladi F. Basch	Norman Fowler	Richard L. Nailen
Carl E. Becker	David C. Griffith	Charles D. Potts
James A. Canham	Thomas M. Gruzs	Elliot Rappaport
Edward G. Cantwell	Kenneth B. Keels	Lynn Saunders
Jane M. Clemmensen	Don O. Koval	Richard E. Singer
Dennis Darling	Alexander McEachern	Anthony W. St. John
Robert J. Deaton	Allen Morinec	David B. Vannoy
William E. Dewitt	William J. Moylan	Raymond M. Waggoner
Thomas W. Diliberti		Donald W. Zipse

Former members of the Working Group and others who contributed to the development of this document are as follows:

John E. Curlett	Phillip E. Gannon	Hugh O. Nash
John B. Dagenhart	Joseph J. Humphrey	Pat O'Donnell
John G. Dalton	J. Frederick Kalbach*	Marek J. Samotyj
Michael J. DeMartini	Robert Keis	William M. Smith
Arthur Freund	Emanuel E. Landsman	Clarence P. Tsung
David A. Fuhrman	Ralph H. Lee*	John J. Waterman*
	William A. Moncrief	

*deceased

Special appreciation is expressed to the following companies and organizations for contributing direct financial support to make possible the development of this Recommended Practice:

Basic Measuring Instruments
Best Power Products
Cleveland Electric Illuminating Company
Collective Intelligence
Delmarva Power
Dranetz Technologies
Electric Power Research Institute
Erico Products
San Diego Gas & Electric

The following persons were on the balloting committee that approved this document for submission to the IEEE Standards Board:

Lucas Ananian	James R. Harvey	R. L. Nailen
Robert J. Beaker	Charles R. Heising	William J. Neiswender
James Beall	Robert W. Ingham	Neil Nichols
Carl E. Becker	Gordon S. Johnson	Pat O'Donnell
Gordon Bracey	W. J. Kelly	Dev Paul
Edward C. Cantwell	Thomas S. Key	Don Pomering
Rene Castenschiold	Don Koval	Elliott Rappaport
James M. Daly	Steven A. Larson	Milton D. Robinson
Robert J. Deaton	Richard H. McFadden	Vincent Saporita
Jerry Frank	Robert Medley	Lynn F. Saunders
Phil E. Gannon	Harold Miles	Thomas E. Sparling
Dan Goldberg	Bill J. Moylan	Charles F. Venugopolan
Shan M. Griffith		Donald W. Zipse

When the IEEE Standards Board approved this standard on June 18, 1992, it had the following membership:

Marco W. Migliaro, *Chair*   **Donald C. Loughry,** *Vice Chair*
Andrew G. Salem, *Secretary*

Dennis Bodson	Donald N. Heirman	T. Don Michael*
Paul L. Borrill	Ben C. Johnson	John L. Rankine
Clyde Camp	Walter J. Karplus	Wallace S. Read
Donald C. Fleckenstein	Ivor N. Knight	Ronald H. Reimer
Jay Forster*	Joseph Koepfinger*	Gary S. Robinson
David F. Franklin	Irving Kolodny	Martin V. Schneider
Ramiro Garcia	D. N. "Jim" Logothetis	Terrance R. Whittemore
Thomas L. Hannan	Lawrence V. McCall	Donald W. Zipse

*Member Emeritus

Also included are the following nonvoting IEEE Standards Board liaisons:

Satish K. Aggarwal
James Beall
Richard B. Engelman
David E. Soffrin
Stanley Warshaw

Paula M. Kelty
IEEE Standards Project Editor

IEEE Recommended Practice for Powering and Grounding Sensitive Electronic Equipment

Thomas S. Key, *Working Group Chair*
Warren H. Lewis, *Secretary;* **Van Wagner,** *Secretary*
Michael C. Keeling, *Technical Editor*
François D. Martzloff, *Coordinating Editor*

This Recommended Practice was developed by a team involving the Working Group members and other contributors listed in the foreword. This team included specialists from electric utilities, load equipment manufacturers, architects and engineers, independent consultants, facility engineers, academia, power conditioning and monitoring equipment manufacturers, and several IEEE Societies. The work was coordinated by Chapter Chairs as follows:

Chapter 1— Introduction:
Thomas S. Key

Chapter 2— Definitions:
Carl E. Becker

Chapter 3— General Needs Guidelines:
François D. Martzloff

Chapter 4— Fundamentals:
J. Frederick Kalbach; Michael C. Keeling;
Raymond M. Waggoner

Chapter 5— Instrumentation:
Jane M. Clemmensen

Chapter 6— Site Surveys and Site Power Analyses:
Edward G. Cantwell

Chapter 7— Case Histories:
Donald W. Zipse

Chapter 8— Specification and Selection of Equipment and Materials:
Vladi F. Basch; John E. Curlett; John J. Waterman

Chapter 9— Recommended Design and Installation Practices:
Warren H. Lewis

Contents

SECTION	PAGE

1. Introduction 21
   1.1 Scope 21
   1.2 Background 21
   1.3 Road Map 23

2. Definitions 25
   2.1 Introduction 25
   2.2 Alphabetical Listing of Terms Used in This Recommended Practice 25
   2.3 Word Avoided Because of No Single Technical Definition 34
   2.4 Abbreviations and Acronyms 35
   2.5 References 36
   2.6 Bibliography 37

3. General Needs Guidelines 39
   3.1 Introduction 39
       3.1.1 Historical Perspective 39
       3.1.2 Proliferation of Disturbing Loads 40
       3.1.3 The Concept of Power Quality 40
       3.1.4 Conflicting Design Philosophies and Safety 41
   3.2 Power Quality Considerations 41
       3.2.1 General Discussion 41
       3.2.2 Classification of Disturbances 42
       3.2.3 Origin of Disturbances 42
       3.2.4 Power Quality Site Surveys 45
   3.3 Grounding Considerations 46
   3.4 Protection Against Disturbances 47
       3.4.1 General 47
       3.4.2 Noise Protection 48
       3.4.3 Surge Protection 49
       3.4.4 Sag Protection 49
   3.5 Safety Systems 50
   3.6 Information Technology Systems 50
   3.7 Shielded, Enclosed EMI/EMC Areas (TEMPEST) 51
       3.7.1 General 51
       3.7.2 Electrical Safety Requirements 51
       3.7.3 Basic Requirements 52
   3.8 Coordination with Other Codes, Standards, and Agencies 52
       3.8.1 General 52
       3.8.2 National Electrical Code (NEC) 52
       3.8.3 Underwriters Laboratories (UL) Standards 53
       3.8.4 Other Laboratories and Testing Agencies 53

SECTION			PAGE

		3.8.5	National Electrical Manufacturers Assocation (NEMA) Standards 53
		3.8.6	National Institute of Standards and Technology (NIST) 53
		3.8.7	International Standards 53
	3.9	References 54	
	3.10	Bibliography 55	

4. Fundamentals 57
    4.1   Introduction 57
    4.2   Impedance Considerations 57
            4.2.1   Frequencies of Interest 57
            4.2.2   Power Source Impedance 58
            4.2.3   Building AC Distribution System Impedance 61
            4.2.4   Load Impedance 61
            4.2.5   Resonance Considerations 64
    4.3   Utility Level Distribution Voltage Disturbances 68
    4.4   Load and Power Source Interactions 68
            4.4.1   Transient Voltage Disturbance Sources/Characteristics 71
            4.4.2   Potential Impacts of Transient Voltage Disturbances ... 72
            4.4.3   Steady-state Voltage Distortion Sources/Characteristics. 73
            4.4.4   Potential Impacts of Steady-state Current Distortions 74
            4.4.5   Corrective Means 77
    4.5   Voltage Surges 77
            4.5.1   Sources/Characteristics 78
            4.5.2   Coupling Mechanisms 84
            4.5.3   Interaction With Buried Cables 86
            4.5.4   Interaction With Above-Ground Conductors 87
            4.5.5   Potential Impact 87
            4.5.6   Surge Voltage Frequency 88
    4.6   Grounding Systems 90
            4.6.1   Earth Electrode Subsystem 90
            4.6.2   Grounding for Fault/Personnel Protection Subsystem ... 90
            4.6.3   Signal Reference Subsystem 91
            4.6.4   Lightning Protection Subsystem 93
    4.7   Shielding Concepts 94
            4.7.1   Electrostatic Shielding 94
            4.7.2   Electromagnetic Shielding 95
    4.8   Bonding Concepts 95
    4.9   References 96
    4.10 Bibliography 97

5. Instrumentation 103
    5.1   General Discussion 103
    5.2   Wiring and Grounding Measurement Instruments 103
    5.3   Infrared Detector 106

SECTION		PAGE
5.4	Root-Mean-Square (RMS) Voltmeters	106
5.5	True RMS Voltmeters	106
	5.5.1 Thermocouple Type	106
	5.5.2 Square-Law Type	107
	5.5.3 Sampling Device	107
5.6	Direct-Reading Ammeters	107
5.7	True RMS Ammeters	107
	5.7.1 Current-Transformer (CT) Ammeters	107
	5.7.2 Hall-Effect Ammeters	108
5.8	Current Measurement Considerations	109
	5.8.1 DC Component on AC Current	109
	5.8.2 Steady-state Values	109
	5.8.3 Inrush and Start-up Current Values	109
	5.8.4 Crest Factor	109
5.9	Receptacle Circuit Testers	109
5.10	Ground Circuit Impedance Testers	109
5.11	Earth Ground Resistance Testers	110
5.12	Oscilloscope Measurements	111
	5.12.1 Line Decoupler and Voltage Measurements	111
	5.12.2 Clamp-on Current Transducer and Current Measurements	112
5.13	Power Disturbance Monitors	112
	5.13.1 Event Indicators	113
	5.13.2 Text Monitors	114
	5.13.3 Waveform Analyzers	117
5.14	Spectrum Analyzers and Computer-Based Harmonic Analysis	120
5.15	Expert Systems	120
	5.15.1 Data Collection Techniques	120
	5.15.2 Recording and Reporting Mechanicsms	120
	5.15.3 Analysis Functions	121
5.16	Electrostatic Discharge	122
5.17	Radio Frequency Interference (RFI) and Electromagnetic Interference (EMI)	122
5.18	Temperature and Relative Humidity	122
5.19	References	122
5.20	Bibliography	122
6. Site Surveys and Site Power Analyses		123
6.1	Introduction	123
6.2	Objectives and Approaches	123
6.3	Coordinating Involved Parties	124
	6.3.1 The Equipment User, Owner, or Customer	124
	6.3.2 The Electronic Equipment Manufacturer/Supplier	124
	6.3.3 The Independent Consultant	124
	6.3.4 The Electrical Contractor or Facility Electrician	125

SECTION			PAGE
	6.3.5	The Electric Utility Company	125
6.4	Conducting a Site Survey		125
	6.4.1	Condition of the Power Distribution and Grounding System	126
	6.4.2	Quality of AC Voltage	136
	6.4.3	Electronic Equipment Environment	140
6.5	Applying Data To Select Cost-Effective Solutions		141
6.6	Long-term Power Monitoring		141
6.7	Conclusions		142
6.8	References		142
6.9	Bibliography		142
7.	Case Histories		145
7.1	General Discussion		145
7.2	Typical Utility-Sourced Power Quality Problems		145
	7.2.1	Voltage Sags Due to Utility Fault Clearing	145
	7.2.2	Voltage Surges Due to Utility Power-Factor/Voltage-Regulation Capacitor Switching	146
7.3	Premises Switching Generated Surges		147
7.4	Electronic Loads		149
	7.4.1	Uninterruptible Power Supply—Unfiltered Input	149
	7.4.2	Uninterruptible Power Supply—Unfiltered Output	149
	7.4.3	Automated Office	150
	7.4.4	Interaction Between Power-Factor/Voltage-Regulation Capacitors and Electronic Loads	151
7.5	Premises-Wiring-Related Problems		151
	7.5.1	Receptacle Level Miswiring	152
	7.5.2	Feeder and Branch Circuit Level Miswiring	152
	7.5.3	Ground-Fault Circuit Interrupter Problems	152
	7.5.4	Ground Discontinuity	152
7.6	Transient Voltage Surge Suppression Network Design		154
7.7	Typical Radiated EMI Problems		155
7.8	Typical Electrical Inspection Problems		155
7.9	Typical Life-Safety System Problems		156
7.10	Typical Misapplication of Equipment Problems		156
7.11	References		157
8.	Specification and Selection of Equipment and Materials		159
8.1	General		159
8.2	Commonly Used Power-Enhancement Devices		160
	8.2.1	Isolation Transformers	161
	8.2.2	Noise Filters	164
	8.2.3	Harmonic Current Filters	164
	8.2.4	Surge Suppressors	165
	8.2.5	Voltage Regulators	166

SECTION			PAGE
	8.2.6	Power Line Conditioners	169
	8.2.7	Computer Power Centers	172
	8.2.8	Standby Power Systems	172
	8.2.9	Uninterruptible Power Supplies (UPS)	174
8.3	Equipment Procurement Specifications		177
	8.3.1	Facility Planner's Considerations	178
	8.3.2	Reliability Considerations	181
	8.3.3	Installation Cost Considerations	185
	8.3.4	Cost of Operation Considerations	186
	8.3.5	Specifying Engineer's Considerations	187
	8.3.6	Transfer Characteristics	189
	8.3.7	Power Technology Considerations	189
8.4	Equipment and Material Specifications		190
	8.4.1	General Discussion	191
	8.4.2	Using Vendor-Supplied Specifications	191
	8.4.3	Creative Specifications	191
	8.4.4	"Mixed" Vendor Specifications	192
	8.4.5	Generic Specifications for Multiple Vendors	193
8.5	Verification Testing		193
	8.5.1	Visual Inspection	193
	8.5.2	Load Tests	194
	8.5.3	Transfer Tests	194
	8.5.4	Synchronization Tests	194
	8.5.5	AC Input Failure and Return Test	194
	8.5.6	Efficiency Test	194
	8.5.7	Load Performance Test	194
	8.5.8	Load Imbalance Test	195
	8.5.9	Overload Capability Test	195
	8.5.10	Harmonic Component Test	195
8.6	Equipment Maintenance		195
	8.6.1	Preventative Maintenance	195
	8.6.2	Wear and Aging of Components	195
	8.6.3	Restoring System Operation After Failure	196
8.7	Bibliography		197
9. Recommended Design/Installation Practices			199
9.1	General Discussion		199
	9.1.1	Safety	199
	9.1.2	Performance	200
	9.1.3	Three-Phase Versus Single-Phase Systems and Loads	200
	9.1.4	Selection of System Voltages	200
	9.1.5	AC System Waveforms	201
9.2	Computer Room Wiring and Grounding		201
	9.2.1	ANSI/NFPA 75-1992	201
	9.2.2	UL 1950-1989	201

SECTION			PAGE

9.3 Dedicated and Shared Circuits 202
    9.3.1 Typical Forms of Unwanted Interaction on Shared Circuits..202
    9.3.2 Dedicated Load Circuits 202
9.4 Feeders 202
    9.4.1 Voltage Drops in Feeders................................202
    9.4.2 Current Ratings for Feeders 204
    9.4.3 Busway Configurations....................................204
9.5 Branch Circuits..204
    9.5.1 Voltage Drop on Branch Circuits....................204
    9.5.2 Shielding of Branch Circuits 204
9.6 Avoiding Single-Phase Input Conditions on Three-Phase Load Equipment 204
9.7 Harmonic Cirrent Control on the AC Supply Wiring System ...204
9.8 Power Factor Improvement..204
9.9 Specialized AC Source Tansfer Switching....................205
9.10 Grounding, General...205
    9.10.1 Annotating Mechanical and Electrical Drawings.......205
    9.10.2 Solidly Grounded AC Supply Systems.....................205
    9.10.3 Working with Improperly Grounding Equipment.......205
    9.10.4 Grounding of Building Structural Steel...................205
    9.10.5 Buried Ring-Ground Electrode System....................206
    9.10.6 Bonding Across Building's Interior/Exterior Line of Demarcation and a Buried Ring-Ground................206
    9.10.7 Grounding Mechanical Equipment in Electronic Areas 206
    9.10.8 Associated Electrical Conduits/Raceways and Enclosures 206
    9.10.9 Bonding Across Exansion Joints in Electrical Conduits...207
    9.10.10 Grounding of AC Services and Systems 207
    9.10.11 Grounding of Separately Derived AC Sources...........207
    9.10.12 Isolated/Insulated Ground (IG) Method 207
    9.10.13 High-Frequency (HF) Ground Referencing Systems...210
    9.10.14 Earth Grounding Electrode Resisteance and Impedance 216
    9.10.15 Earth Grounding Electrode Conductor Resistance and Impedance 216
    9.10.16 Recommended Grounding of Communication Systems 216
    9.10.17 AC System Grounding for Uninterruptible Power Supply (UPS) With Bypass Circuits............................217
    9.10.18 Grounding Air Terminals (Lightning Rods)............217
    9.10.19 Galvanized Construction Channel as a Grounding Bus-Bar..217
    9.10.20 Uninsulated Grounding Conductors........................218

SECTION		PAGE
9.11	Lightning/Surge Protection	218
9.11.1	Service Entrance Lightning/Surge Protection	218
9.11.2	Premise Electrical System Lightning/Surge Protection	219
9.11.3	UPS Surge Protection	220
9.11.4	Data (Communications and Control) Cabling and Equipment Lightning/Surge Protection	220
9.11.5	Telecommunication System Lightning/Surge Protection	221
9.12	400 Hz (380–480 Hz) Power Systems	225
9.12.1	Recommended Location of the 400 Hz AC System	226
9.12.2	General Grounding and Shielding of 400 Hz Systems	226
9.12.3	Controlling 400 Hz Wiring Losses	226
9.12.4	400 Hz Conductor Ampacity	228
9.12.5	Component Derating at 400 Hz	228
9.13	Switchboards	229
9.13.1	Grounding Bus-Bars	229
9.13.2	Neutral Bus-Bars	229
9.14	Panelboards	229
9.14.1	Recommended Line Bus-Bar Ampacity	229
9.14.2	Recommended Neutral Bus-Bar Ampacity and Wiring Capacity	230
9.14.3	Equipment Ground Bus-Bar Ampacity and Wiring Capacity	230
9.14.4	Panelboard Mounting and Grounding	230
9.14.5	Location of the Panelboard	230
9.15	Power Distribution Units (PDUs) for AC Power-Load Interface	231
9.16	Automatic Line Voltage Regulating Transformers (ALVRTs)	231
9.17	Dry-Type Tansformer	231
9.17.1	Electrostatic Shielding in the Transformer	231
9.17.2	Nonlinear Load Impact on Transformers	231
9.17.3	Neutral Connection Derating for the Wye-Connected Transformer	235
9.17.4	Transformer Percent Impedance	235
9.17.5	Transformer Forward-Transfer Impedance	235
9.17.6	Banked Transformers, Open Deltas, Tees, and Common Cores	235
9.17.7	Making Conduit/Raceway Connections to Transformers	236
9.17.8	Separation of Input from Output Wiring on Transformers	236
9.18	Wiring Devices	236
9.18.1	NEMA, IEC, and Other Configurations	236
9.18.2	Wiring Device Conductor Terminations	236
9.18.3	Special Keying for Non-60 Hz Circuits	237
9.18.4	Grounding Configurations and Requirements	237

SECTION			PAGE
	9.18.5	Neutral (Grounded Conductor) Configurations and Requirements	237
9.19	Pull and Junction Boxes		237
9.20	Metal Conduit and Metal Enclosed Wireway		237
	9.20.1	Recommended Materials	237
	9.20.2	Metal Conduit for Signal Conductors	237
	9.20.3	Conduit Couplings	238
	9.20.4	End-Terminating Fittings with Locknuts on Conduits	238
	9.20.5	End-Terminating Fitting Use With Concentric Knockouts	238
	9.20.6	Special Requirements for Liquid-tight Terminating Fittings	238
	9.20.7	Use of Terminating Fittings With Reducing Washers	238
9.21	References		239
9.22	Bibliography		240

FIGURES

Fig 2-1	Distortion Example	28
Fig 2-2	Interruption	29
Fig 2-3	Noise Example	30
Fig 2-4	Notches	31
Fig 2-5	Sag	32
Fig 2-6	Swells Occurring Upon Recovery From a Remote System Fault	33
Fig 3-1	Sources of Disturbances (a) Examples of Interferences That Do Not Involve Utility Supply (b) Example of Source Interference	44
Fig 3-2	Typical Design Goals of Power-Conscious Computer Manufacturers	47
Fig 3-3	Relationship Between Amplitude, Duration, Rate of Change of Disturbances, and Their Effects on Equipment	48
Fig 4-1	First Order Model of Transformer Impedances	58
Fig 4-2	(a) Phase-Neutral Transient Resulting From Addition of Capacitive Load (b) Neutral-Ground Transient Resulting From Addition of Inductive Load	60
Fig 4-3	Phase-Neutral Transient Resulting From Arcing and Bouncing Contactor	61
Fig 4-4	Typical AC Distribution Branch Circuit Impedance Versus Frequency	62
Fig 4-5	(a) Passive Load Resistance Versus Frequency	63
	(b) Passive Load Inductive Reactance Versus Frequency	63
	(c) Passive Load Capacitive Reactance Versus Frequency	64
Fig 4-6	(a) Series R-L-C Circuit Impedance Versus Frequency	65
	(b) Parallel R-L-C Circuit Impedance Versus Frequency	66
Fig 4-7	Resonance Characteristics of Conductors	67
Fig 4-8	Generalized Power Network	78
Fig 4-9	Typical AC Building Distribution Wiring System	79

FIGURES		PAGE
Fig 4-10	(a) Typical Behavior of (Power-off) Switching Transient (Recovery) Voltage Without Multiple Interruption-Reignition	80
	(b) Typical Behavior of (Power-off) Switching Transient (Recovery) Current Without Multiple Interrruption-Reignition.	80
Fig 4-11	Distribution of Lightning Stroke Current	81
Fig 4-12	Arcing Distances for Bare and Insulated Conductors	83
Fig 4-13	Inductive Coupling of Surge Currents to Adjacent Sensitive Circuits	85
Fig 4-14	Normalized Induced Voltage Into Sensitive Circuits	85
Fig 4-15	Frequency Spectra of Common Surge Test Waveforms	89
Fig 4-16	Grounding for Fault/Personnel Protection Subsystem	90
Fig 4-17	Combined Safety and Signal Reference Grounding Subsystems	91
Fig 4-18	Residual Surge Voltage Versus Frequency for 100 kHz Ring Wave	93
Fig 4-19	Electrostatic Field Between Charged Conductors	95
Fig 5-1	Multimeters With Clamp-on Current Probes	108
Fig 5-2	Ground Circuit Impedance Testers	110
Fig 5-3	Voltage Measurements Can Be Made With Oscilloscopes Using Line Attenuators	111
Fig 5-4	Text Monitors Can Be Used to Detect AC RMS Voltage Variations	115
Fig 5-5	Text Monitors Are Typically Lightweight and Well-Suited to Use in the Field	115
Fig 5-6	The Newest Category of Text Monitors is Capable of Making Simultaneous Voltage and Current Measurements	116
Fig 5-7	Waveform Analyzers Detect, Capture, and Record Power Line Disturbance Events as Complete Waveforms	118
Fig 5-8	Data Output From Waveform Analyzers	118
Fig 5-9	Some Waveform Analyzers Are Capable of Down-loading Data to a Personal Computer	119
Fig 5-10	Expert Systems Use Data Input by the User, Software Procedures, and Rules	121
Fig 6-1	Sample Set of Forms	131
Fig 6-2	Recommended Power Monitor Hookup Procedure for Single-Phase Applications	136
Fig 6-3	Recommended Power Monitor Hookup for Single-Phase With Power Conditioner	137
Fig 6-4	Recommended Power Monitor Hookup for Three-Phase Wye	137
Fig 8-1	Summary of Performance Features for Various Types of Power Conditioning Equipment	162
Fig 8-2	Isolation Transformer	163
Fig 8-3	Shielded Isolation Transformer	163
Fig 8-4	LC Filter	164
Fig 8-5	Harmonic Current Filter	165
Fig 8-6	Tap Changing Regulator	167

FIGURES		PAGE
Fig 8-7	Buck-Boost Regulator	167
Fig 8-8	Ferroresonant Regulator	168
Fig 8-9	Magnetic Synthesizer	169
Fig 8-10	Motor-Generator Set	170
Fig 8-11	Standby Power System	173
Fig 8-12	Rotary UPS With DC Motor/Generator	174
Fig 8-13	Rotary UPS With Rectifier/DC Motor	175
Fig 8-14	Rotary UPS With Inverter	175
Fig 8-15	(a) Rectifier/Charger UPS	176
	(b) Line Interactive UPS	177
Fig 8-16	Parallel Redundant System	182
Fig 8-17	Isolated Redundant UPS	183
Fig 9-1	Recommended Separation of Sensitive Equipment Power Distribution From Support Equipment Power Distribution	203
Fig 9-2	Best Design Locates Shielded Isolation Transformer as Close to Sensitive Loads as Possible	208
Fig 9-3	(a) Isolated Grounding Conductor Pass-Through Distribution Panel (b) Isolated Grounding Conductor Wiring Method With Separately Derived Source	211
Fig 9-4	Raised Access Flooring Substructure as Signal Reference Grid	213
Fig 9-5	Signal Reference Grid Fabricated From Copper Strips	214
Fig 9-6	Typical Locations of Power Distribution TVSS	219
Fig 9-7	Recommended Installation Approach for Signal Protectors	222
Fig 9-8	Recommended Installation Practice for Combined Power TVSS and Signal Protectors	223
Fig 9-9	Signal Reference Potential Difference Created by Diverting Transient Energy Through Grounding Conductor of Portion of System	224
Fig 9-10	Protection Against Reference Voltage Differences Caused by Two-Port Connection	224
Fig 9-11	Transformer Capability for Supplying Electronic Loads	232
Fig 9-12	Example Distribution of Harmonic Levels in a Facility AC Distribution System	234

TABLES

Table 4-1	Example Cable Impedance at High Frequencies	
	(a) #4 AWG Building Wire (25 mm^2)	62
	(b) #4/0 AWG Building Wire (107 mm^2)	62
Table 4-2	Standard Nominal System Voltages and Voltage Ranges	69
Table 4-3	Matching Sensitive Load and Power Source Requirements With Expected Environments	70
Table 4-4	Example Input Harmonic Current Distortion in Balanced Three-Phase Circuits Due to Rectifier-Capacitor Power Supply	74

TABLES		PAGE
Table 4-5	Example Calculation of a Nonlinear Load's K-Factor	76
Table 4-6	Thresholds of Failure of Selected Semiconductors	89
Table 5-1	Recommended Test Instruments for Conducting a Site Survey	104
Table 8-1	Summary of Power-Enhancement Devices	160
Table 9-1	415 Hz Impedance in Ohms/100 ft	227

APPENDIX

Interpreting and Applying Existing Power Quality Studies 241

APPENDIX TABLES

Table A1	Comparison of National Power Quality Studies	242
Table A2	Monitor Threshold Settings	242
Table A3	Normal Mode Power Disturbances per Year	243
Table A4	Expected Disruptive Power Disturbances per Year	244
Table A5	Causes of 100 Disruptive Undervoltages	245
Table A6	Disruptive Undervoltages per Year	246

Chapter 1
Introduction

1.1 Scope. This Recommended Practice presents recommended engineering principles and practices for powering and grounding sensitive electronic equipment.

The scope and purpose of this document are limited to recommended design, installation, and maintenance practices for electrical power and grounding (including both power-related and signal-related noise control) of sensitive electronic processing equipment used in commercial and industrial applications.

The main objective is to provide a consensus of recommended practices in an area where conflicting information and confusion, stemming primarily from different view points of the same problem, have dominated. Practices herein address electronic equipment performance issues while maintaining a safe installation, as specified in the National Electrical Code (NEC) and recognized testing laboratories' standards. This Recommended Practice is not intended to replace or to take precedence over any codes or standards adopted by the jurisdiction where the installation resides.

1.2 Background. As sensitive electronic loads proliferate in industrial and commercial power systems, so do problems related to power quality. Powering and grounding sensitive electronic equipment has been a growing concern for commercial and industrial power system designers. This concern frequently materializes after start-up, when electronic system operating problems begin to occur. Efforts to alleviate these problems have ranged from installing expensive power conditioning equipment to applying special grounding techniques that are not found in conventional safe grounding practice. Grasping for conditioning equipment or "magic" grounding methods is a common response. In some cases this approach has led to unsafe practices and violations of the NEC, without solving operating problems. In response to this situation, this Recommended Practice attempts to provide a true understanding of the fundamentals of powering and grounding sensitive electronic equipment and the various types of problems that can arise—knowledge that has been lacking in commercial and industrial power systems.

The concept of load and source compatibility is not new. The need to provide power with steady voltage and frequency has been recognized since the inception of the electric utility industry. However, the definition of "steady" has changed over the years, reflecting the greater susceptibility of increasingly sophisticated electronic equipment to the departure from "steady" conditions.

Some of the early concerns were flicker of light bulbs due to voltage variations, and overheating of electromagnetic loads or interference of communication loads due to voltage waveform distortion. Recognition of these problems led to the development of voluntary standards that contributed significantly to reducing occurrences.

More recently, transient voltage disturbances associated with short circuits, lightning, and power system switching have emerged as a major concern to manufacturers and users of electronic equipment. The issue of grounding, and particularly how to deal with noise and safety simultaneously, is complicated by conflicting philosophies advocated by people of different backgrounds. Power-oriented engineers and signal-oriented engineers often differ in their perception of common problems and solutions.

Since the earliest days of electric power, users have desired that utilities provide electricity without interruptions, surges, and harmonic waveform distortions. Reducing such power line disturbances has always been a critical concern for utilities. Recently, however, new sources of disturbances have begun to proliferate, just as many pieces of equipment are becoming more sensitive to these same power disturbances. Some of these disturbances are generated by adjacent equipment and by inadequate wiring and grounding practices. These developments have presented utilities and users with a new set of complex power quality issues that require wide-reaching cooperative efforts to be resolved.

Today's complaints about the quality of power are not easily resolved because they involve both a multitude of different causes and a variety of specific sensitivities in the affected equipment. A commonly applied solution to power incompatibilities is to install interface equipment between commercial power and sensitive loads. Difficulties in assessing this need are as follows:

(1) The quantification of precisely how much downtime is power-related, and
(2) The subjective nature of estimating the cost of sensitive load misoperation that is attributable to power line disturbance.

The cost/benefit aspects of the problem can be addressed from a technical point of view in standards, but detailed economic analysis and specific decisions remain the prerogative of the user. Power system designers; utility companies; and manufacturers of sensitive electronic equipment, potential sources of interference, and voltage conditioning equipment must cooperate with each other to find effective solutions.

As in the past, voluntary consensus standards are also needed. Focusing on the technical issues, dispelling misconceptions, and recommending sound practices can assist the user in making informed economic decisions. One of the goals of this Recommended Practice is to promote a better understanding of the significant issues and to dispel misconceptions.

Fortunately, powering and grounding an electronic system is fundamentally the same as any electrical system. Estimating the load, matching current and voltage requirements, or planning for future growth involves the same basic information. Similarly, designing an appropriate electrical distribution system, selecting and coordinating overcurrent protection, and assuring voltage regulation, makes use of the same engineering practices. Even the principles of grounding for safety can be applied to electronic loads in the same way as to any other load.

The IEEE Color Book series is an excellent reference library available for designing commercial and industrial power systems of all types. There are currently ten Color Books in the series, each of which provides recommended practices in a specific subject area. The objective is to assist in the design of safe, reliable, and economical electric power systems by providing the consensus of knowledge and experience of the contributing IEEE members. The Emerald Book is a new entry in the series, directed specifically at powering and grounding sensitive electronic equipment.

1.3 Road Map. The following provides the reader with a road map of this Recommended Practice.

Chapter 2 provides definitions of the descriptors that pertain specifically to power quality issues and are generally not otherwise available in IEEE standards. Also provided is a list of terms that have been deliberately avoided in this Recommended Practice because they have a wide range of different meanings and no generally accepted single technical definition.

General need guidelines are provided in Chapter 3. This chapter is intended to identity the relevant codes and standards as well as the existing electrical environments to which equipment is typically subjected. These guidelines are established as a basis for the treatment of performance definition and recommended practices in subsequent chapters.

Chapter 4 introduces the reader to the fundamental concepts necessary for understanding and applying recommended practices. Fundamentals not unique to sensitive electronic and electrical equipment are treated very lightly, or by reference to other standards.

Chapter 5 presents information on available measurement instruments that are needed to investigate and diagnose problems in power systems that serve sensitive electronic equipment.

Site power analysis and site surveys are covered in Chapter 6. This chapter draws from the technical information base established in previous chapters, and presents practical applications of problem identification and diagnosis. Measurement techniques are also discussed. The approach is to start with wiring and grounding checks and progress through voltage amplitude and disturbance measurements to harmonic analysis. Electromagnetic interference (EMI) is covered primarily by reference.

Case histories are presented in Chapter 7. These are intended to provide examples of real-world performance and safety problems that have been encountered in the field. Cases are selected that illustrate the need to follow specific recommended practices and potential results when recommended practices are not followed.

Chapter 8 presents the myriad of available power enhancement equipment from the points of view of basic technology, performance, and function. Performance verification is also covered.

Chapter 9 covers both recommended and nonrecommended design and installation practices. The intent is to present the Working Group's collective engineering experience and judgment of effective practices. The discussion of nonrecommended practices very deliberately attacks "old wives' tales" and raises "warning flags," but does not introduce any new concepts or materials. The discouragement of commonly observed nonrecommended practices is emphasized.

Chapter 2
Definitions

2.1 Introduction. The electronic power community is pervaded by terms that have no scientific definition; one of the purposes of this chapter is to eliminate the use of those words. Another purpose of this chapter is to define those terms that will aid in the understanding of concepts within this Recommended Practice. Where possible, definitions were obtained from the IEEE Std 100-1988 [4].[1] The second choice was to use other appropriate sources, and the final choice was to use a new definition that conveys a common understanding for the word as used in the context of this Recommended Practice.

This chapter is divided into three parts. First, an alphabetical listing of definitions is provided in 2.2. The reader is referred to the IEEE Std 100-1988 for all words not listed herein. The second subsection (2.3) lists those terms that have been deliberately avoided in this document because of no generally accepted single technical definition. These words find common use in discussing ac distribution-related power problems, but tend not to convey significant technical meaning. The third subsection (2.4) lists abbreviations that are employed throughout this Recommended Practice.

2.2 Alphabetical Listing of Terms Used in This Recommended Practice. The primary source for the definitions in this section is IEEE Std 100-1988 [4].

This section does not include any device or equipment definitions (e.g., isolation transformers, uninterruptible power systems); the reader is advised to refer to the index. Most pertinent equipment is described in Chapter 8.

bonding. (1) The electrical interconnecting of conductive parts, designed to maintain a common electrical potential [2]. (2) The permanent joining of metallic parts to form an electrically conductive path which will assure electrical continuity and the capacity to conduct safely any current likely to be imposed [2].

commercial power. Electrical power furnished by the electric power utility company.

common-mode noise. The noise voltage that appears equally and in phase from each current-carrying conductor to ground.

NOTE: For the purposes of this Recommended Practice, this definition expands the existing definition in IEEE Std 100-1988 [4] (previously given only for signal cables) to the power conductors supplying sensitive sensitive electronic equipment.

---

[1] The numbers in brackets correspond to those of the references in 2.5; when preceded by the letter "B," they correspond to those in the bibliography in 2.6.

coupling. Circuit element or elements, or network, that may be considered common to the input mesh and the output mesh and through which energy may be transferred from one to the other [4].

crest factor. Ratio between the peak value (crest) and rms value of a periodic waveform [4].

critical load. Devices and equipment whose failure to operate satisfactorily jeopardizes the health or safety of personnel, and/or results in loss of function, financial loss, or damage to property deemed critical by the user.

degradation failure. *See:* **failure, degradation.**

differential-mode noise. *See:* **noise, transverse-mode.**

direct-reading ammeters. Ammeters that are employed with a series shunt and that carry some of the line current through them for measurement purposes. They are part of the circuit being measured.

displacement power factor. *See:* **power factor, displacement.**

distortion factor. The ratio of the root-mean-square of the harmonic content to the root-mean-square value of the fundamental quantity, expressed as a percent of the fundamental [4]. Also referred to as *total harmonic distortion* [6].

dropout. A loss of equipment operation (discrete data signals) due to noise, sag, or interruption.

dropout voltage. The voltage at which a device will release to its deenergized position (for this Recommended Practice, the voltage at which a device fails to operate).

efficiency (of a power system). The relationship between the input power that a power system draws and the corresponding power that it is able to supply to the load (kilowatt out/kilowatt in).

equipment grounding conductor. The conductor used to connect the noncurrent carrying parts of conduits, raceways, and equipment enclosures to the grounding electrode at the service equipment (main panel) or secondary of a separately derived system (e.g., isolation transformer). (This term in defined more specifically in the NEC [2], Section 100).

failure, degradation. Failure that is both gradual and partial.
NOTE: In time, such a failure may develop into a complete failure [4].

failure mode. The effect by which a failure is observed [4].

flicker. A variation of input voltage sufficient in duration to allow visual observation of a change in electric light source intensity.

form factor (periodic function). The ratio of the root-mean-square value to the average absolute value, averaged over a full period of the function [4].

forward transfer impedance. An attribute similar to internal impedance, but at frequencies other than the nominal (e.g., 60 Hz power frequency). Knowledge of the forward transfer impedance allows the designer to assess the capability of the power source to provide load current (at the harmonic frequencies) needed to preserve a good output voltage waveform. Generally, the frequency range of interest is 60 Hz to 3 kHz, for 5–60 Hz power systems, and 20–25 kHz for 380–480 Hz power systems.

frequency deviation. An increase or decrease in the power frequency. The duration of a frequency deviation can be from several cycles to several hours.

ground. A conducting connection, whether intentional or accidental, by which an electric circuit or equipment is connected to the earth, or to some conducting body of relatively large extent that serves in place of the earth.

NOTE: It is used for establishing and maintaining the potential of the earth (or of the conducting body) or approximately that potential, on conductors connected to it, and for conducting ground currents to and from earth (or the conducting body) [4].

ground electrode. A conductor or group of conductors in intimate contact with the earth for the purpose of providing a connection with the ground [2].

ground electrode, concrete-encased. A grounding electrode completely encased within concrete, located within, and near the bottom of, a concrete foundation or footing or pad, that is in direct contact with the earth. (This term is defined more specifically in Article 250 of the NEC [2].)

ground grid. A system of interconnected bare conductors arranged in a pattern over a specified area and buried below the surface of the earth. (The primary purpose of the ground grid is to provide safety for workmen by limiting potential differences within its perimeter to safe levels in case of high currents which could flow if the circuit being worked became energized for any reason or if an adjacent energized circuit faulted. Metallic surface mats and gratings are sometimes utilized for the same purpose [4].) This term should not be used when referring to a **signal reference structure**, which is defined in this chapter.

ground, high-frequency reference. *See:* **signal reference structure**.

ground impedance tester. A multifunctional instrument designed to detect certain types of wiring and grounding problems in low-voltage power distribution systems.

ground loop. A potentially detrimental loop formed when two or more points in an electrical system that are nominally at ground potential are connected by a conducting path such that either or both points are not at the same ground potential [4].

ground, radial. A conductor connection by which separate electrical circuits or equipment are connected to earth at one point. Sometimes referred to as a star ground.

ground, ufer. *See:* **ground electrode, concrete-encased**.

ground window. The area through which all grounding conductors, including metallic raceways, enter a specific area. It is often used in communications systems through which the building grounding system is connected to an area that would otherwise have no grounding connection.

harmonic distortion. The mathematical representation of the distortion of the pure sine waveform. *See:* **distortion factor**. (See Fig 2-1.)

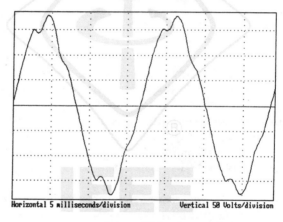

Source: [B2].

**Fig 2-1
Distortion Example**

impulse. *See:* **transient**.

input power factor (of a system). Specifies the ratio of input kilowatts to input kilovoltamperes at rated or specified voltage and load.

input voltage range (of a power system). The range of input voltage that the system can operate over.

DEFINITIONS

inrush. The amount of current that a load draws when it is first turned on.

interruption. The complete loss of voltage for a time period. (See Fig 2-2.)

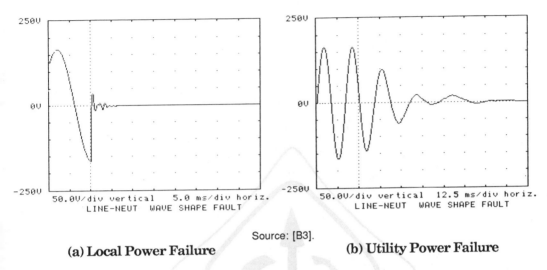

Source: [B3].

(a) Local Power Failure (b) Utility Power Failure

**Fig 2-2
Interruption**

isolated equipment ground. An insulated equipment grounding conductor run in the same conduit or raceway as the supply conductors. This conductor is insulated from the metallic raceway and all ground points throughout its length. It originates at an isolated ground type receptacle or equipment input terminal block and terminates at the point where neutral and ground are bonded at the power source. (This term is defined more specifically in the NEC [2]), Sections 250-74 and 250-75.)

isolation. Separation of one section of a system from undesired influences of other sections.

linear load. An electrical load device which, in steady state operation, presents an essentially constant load impedance to the power source throughout the cycle of applied voltage.

noise. Electrical noise is unwanted electrical signals that produce undesirable effects in the circuits of the control systems in which they occur [4]. (For this Recommended Practice, "control systems" is intended to include sensitive electronic equipment in total or in part.) (See Fig 2-3.)

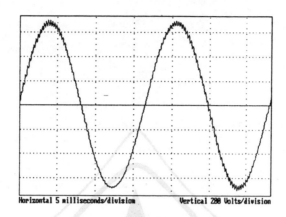

Source: [B2].

Fig 2-3
Noise Example

noise, common-mode. *See:* **common-mode noise.**

noise, differential-mode. *See:* **transverse-mode noise.**

noise, normal-mode. *See:* **transverse-mode noise.**

noise, transverse-mode. *See:* **transverse-mode noise.**

nonlinear load. Electrical load that draws current discontinuously or whose impedance varies during the cycle of the input ac voltage waveform.

nonlinear load current. Load current that is discontinuous or is not proportional to the ac voltage.

notch. A switching (or other) disturbance of the normal power voltage waveform, lasting less than a half-cycle; which is initially of opposite polarity than the waveform, and is thus subtractive from the normal waveform in terms of the peak value of the disturbance voltage. This includes complete loss of voltage for up to a half-cycle. *See:* **transient.** (See Fig 2-4.)

DEFINITIONS

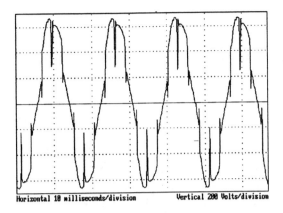

Source: [B2].

**Fig 2-4
Notches**

outage. *See:* **interruption.**

output (reverse transfer) impedance (of a power source). Similar to forward transfer impedance, but it describes the characteristic impedance of the power source as seen from the load, looking back at the source. *See:* **forward transfer impedance.**

overvoltage. A rms increase in the ac voltage, at the power frequency, for durations greater than a few seconds. *See:* **swell** and **surge.**

phase shift. The displacement in time of one periodic-waveform relative to other waveform(s).

power disturbance. Any deviation from the nominal value (or from some selected thresholds based on load tolerance) of the input ac power characteristics.

power disturbance monitor. Instrumentation developed specifically for the analysis of voltage and current measurements.

power factor, displacement. The displacement component of power factor; the ratio of the active power of the fundamental wave, in watts, to the apparent power on the fundamental wave, in volt-amperes [4].

power factor, total. The ratio of the total power input in watts to the total volt-ampere input [4].

power quality. The concept of powering and grounding sensitive electronic equipment in a manner that is suitable to the operation of that equipment.

radial ground. *See:* **ground, radial.**

receptacle circuit tester. A device that, by a pattern of lights, is intended to indicate wiring errors in receptacles. Receptacle circuit testers have some limitations. They may indicate incorrect wiring, but cannot be relied upon to indicate correct wiring.

recovery time. Time interval needed for the output voltage or current to return to a value within the regulation specification after a step load or line change [2]. Also may indicate the time interval required to bring a system back to its operating condition after an interruption or dropout.

recovery voltage. The voltage that occurs across the terminals of a pole of a circuit interrupting device upon interruption of the current [4].

safety ground. *See:* **equipment grounding conductor.**

sag. A rms reduction in the ac voltage, at the power frequency, for durations from a half-cycle to a few seconds. *See:* **notch, undervoltage.** (See Fig 2-5.)
NOTE: The IEC terminology is *dip*.

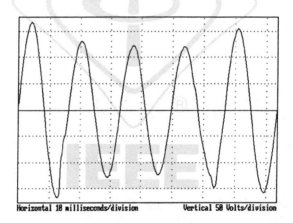

Source: [B2].

**Fig 2-5
Sag**

shield. As normally applied to instrumentation cables, a conductive sheath (usually metallic) applied over the insulation of a conductor or conductors, for the purpose of providing means to reduce coupling between the conductors so shielded and other conductors that may be susceptible to, or that may be generating, unwanted electrostatic or electromagnetic fields (noise).

shielding. Shielding is the use of a conducting barrier between a potentially disturbing noise source and sensitive circuitry. Shields are used to protect cables (data and power) and electronic circuits. They may be in the form of metal barriers, enclosures, or wrappings around source circuits and receiving circuits.

signal reference structure. A system of conductive paths among interconnected equipment that reduces noise-induced voltages to levels that minimize improper operation. Common configurations include grids and planes.

slew rate. Rate of change of (ac voltage) frequency.

star ground. *See:* **ground, radial**.

star-connected circuit. A polyphase circuit in which all the current paths of the circuit extend from a terminal of entry to a common terminal or conductor (which may be the neutral conductor).

NOTE: In a three-phase system this is sometimes called a Y (or wye) connection [4].

surge. *See:* **transient**.

surge reference equalizer. A surge-protective device used for connecting equipment to external systems whereby all conductors connected to the protected load are routed, physically and electrically, through a single enclosure with a shared reference point between the input and output ports of each system.

swell. An rms increase in the ac voltage, at the power frequency, for durations from a half-cycle to a few seconds. *See:* **overvoltage** and **surge**. (See Fig 2-6.)

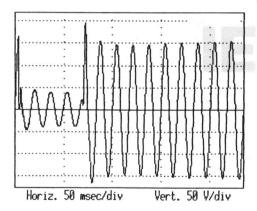 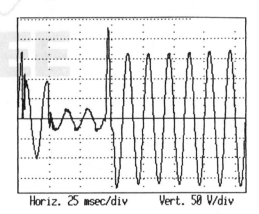

Source: [B1].

**Fig 2-6
Swells Occurring Upon Recovery From a Remote System Fault**

transfer time (uninterruptible power supply). The time that it takes an uninterruptible power supply to transfer the critical load from the output of the inverter to the alternate source or back again.

transient. A subcycle disturbance in the ac waveform that is evidenced by a sharp brief discontinuity of the waveform. May be of either polarity and may be additive to or subtractive from the nominal waveform. *See:* **swell**, **notch**, and **overvoltage**.

transverse-mode noise. (With reference to load device input ac power.) Noise signals measurable between or among active circuit conductors feeding the subject load, but not between the equipment grounding conductor or associated signal reference structure and the active circuit conductors.

unbalanced load regulation. A specification that defines the maximum voltage difference between the three output phases that will occur when the loads on the three are of different levels.

undervoltage. A rms decrease in the ac voltage, at the power frequency, for durations greater than a few seconds.

voltage distortion. Any deviation from the nominal sine waveform of the ac line voltage.

voltage regulation. The degree of control or stability of the rms voltage at the load. Often specified in relation to other parameters, such as input-voltage changes, load changes, or temperature changes.

2.3 Words Avoided Because of No Single Technical Definition. The following words have a varied history of usage, and some may have specific definitions for other applications. It is an objective of this Recommended Practice that the following ambiguous words not be used to generally describe problem areas nor solutions associated with the powering and grounding of sensitive electronic equipment:

blackout
brownout
clean ground
clean power
computer grade ground
conducting barriers
counterpoise ground
dedicated ground
dirty ground
dirty power

equipment safety grounding conductor
frame ground
frequency shift
glitch
natural electrodes
power surge
raw power
raw utility power
shared circuits
shared ground
spike
subcycle outages
Type I, II, III power disturbances

2.4 Abbreviations and Acronyms. The following abbreviations are utilized throughout this Recommended Practice:

AFD	adjustable frequency drive
ALVRT	automatic line voltage regulating transformer
ASAI	average service availability index
CATV	cable accessed television
COTC	central office trunk cable
CPC	computer power center
CPU	central processing unit
CRT	cathode ray tube
CVT	constant voltage transformer
EFT	electrical fast transient
EGC	equipment grounding conductor
EMC	electromagnetic compatibility
EMI	electromagnetic interference
EMT	electrical metallic tubing
ESD	electrostatic discharge
FMC	flexible metal conduit
HF	high frequency
IEC	International Electrotechnical Commission
IG	isolated/insulated grounding
IMC	intermediate metal conduit
IT	isolation transformer
LDC	line drop compensator
M-G	motor-alternator/generator
MCT	metal cable tray

MTBF	mean time between failures
NEMP	nuclear electromagnetic pulse
NEC	National Electrical Code
NIST	National Institute of Standards and Technology
OEM	original equipment manufacturer
OSHA	Occupational Safety and Health Administration
PC	personal computer
PDU	power distribution unit
PLC	power line conditioner
PWM	pulse width modulation
RFI	radio frequency interference
RMC	rigid metal conduit
SDS	separately derived ac system
SE	service entrance
SG	solidly grounded; solid grounding (*See:* **equipment grounding conductor**)
SI	solidly interconnected
SRG	signal reference grid
SRP	signal reference plane
SRS	signal reference structure
THD	total harmonic distortion
TVSS	transient voltage surge suppressor
UL	Underwriters Laboratories
UPS	uninterruptible power supply
VDT	video display terminal

2.5 References. This standard shall be used in conjunction with the following publications. When the following standards are superseded by an approved revision, the revision shall apply:

[1] ANSI C84.1-1989, American National Standard for Electric Power Systems and Equipment—Voltage Ratings (60 Hz).[2]

[2] ANSI/NFPA 70-1993, National Electrical Code.[3]

[3] IEEE Std C62-1990, Complete 1990 Edition: Guides and Standards for Surge Protection. [4]

---

[2] ANSI publications are available from the Sales Department, American National Standards Institute, 11 West 42nd Street, 13th Floor, New York, NY 10036, USA.

[3] NFPA publications are available from Publications Sales, National Fire Protection Association, 1 Batterymarch Park, P.O. Box 9101, Quincy, MA 02269-9101, USA.

[4] IEEE publications are available from the Institute of Electrical and Electronics Engineers, Service Center, 445 Hoes Lane, P.O. Box 1331, Piscataway, NJ 08855-1331, USA.

[4] IEEE Std 100-1988, IEEE Standard Dictionary of Electrical and Electronics Terms (ANSI).

[5] IEEE Std 142-1991, IEEE Recommended Practice for Grounding of Industrial and Commercial Power Systems (IEEE Green Book).

[6] IEEE Std 519-1992, IEEE Guide for Harmonic Control and Reactive Compensation of Static Power Converters (ANSI).

[7] P1159, Recommended Practice on Monitoring Electric Power Quality, D2/June 30, 1992.[5]

2.6 Bibliography

[B1] IEEE Std C62.41-1991, IEEE Recommended Practice on Surge Voltages in Low-Voltage AC Power Systems.

[B2] *The Dranetz Field Handbook for Power Quality Analysis*, Edison, NJ: Dranetz Technologies Inc., 1991.

[B3] McEachern, Alexander, *Handbook of Power Signatures*, Foster City, CA: Basic Measuring Instruments, 1989.

---

[5] This IEEE authorized standards project is available from the Sales Dept., IEEE Service Center, 445 Hoes Lane, Piscataway, NJ 08855-1331.

Chapter 3
General Needs Guidelines

3.1 Introduction. The need to provide reliable power with a steady voltage and frequency has been recognized since the inception of the electric utility industry. However, the engineering reality of a large power system is that disturbances are unavoidable. These disturbances in the quality of power delivery can occur during the normal operation of the system as well as during abnormal conditions, resulting in equipment damage or operational upset. This dichotomy may be the source of misunderstandings at best or disputes at worst between suppliers and users of electric power, and between manufacturers and users of sensitive equipment. One of the goals of this book is to promote better understanding of the significant issues and to dispel some misconceptions about how to avoid or correct problems.

This chapter presents a brief description of the nature of power quality problems, of possible solutions, and of the resources available for assistance in dealing with problems. A brief historical review of the resolution of some of the earlier conflicts provides a useful perspective on solving these types of problems.

3.1.1 Historical Perspective. As public expectations of uniform lighting intensity grew and as more manufacturers began to use electric motors to drive their production lines, utilities adopted stricter standards for voltage regulation. During the 1930s, utilities also found that they had to pay increasing attention to voltage disturbances caused by customer equipment on their distribution lines. Research showed that flicker in incandescent lamps caused by voltage fluctuations could be perceived even if the pulsation on the power line was only a third of a volt on a 120 V system. This type of problem led to an increasing number of industry standards for end-use equipment aimed at reducing voltage fluctuations sent back along a power line.

A somewhat different problem arose during the 1950s as air conditioners rapidly became popular. When early models were switched on, so much energy was used to get their compressors started that the incoming line voltage was temporarily reduced and the motors often could not reach operating speed, ran poorly, or stalled. Fortunately, in this case, a remedy was readily available: adding power-factor correction capacitors in the system.

The reason why today's complaints about the quality of power cannot be handled so simply is that they seem to reflect both a multitude of different causes and a variety of specific sensitivities in the customer equipment most affected. Just as the air conditioner problems were eventually solved by a coordinated effort among affected parties, so too can new standards on equipment and on levels of permissible voltage distortions help guide the design and application of both sensitive electronic equipment and heavy-duty apparatus. Such standards will have to be applied much more selectively than in the past, however, and address a much more complex set of issues.

3.1.2 Proliferation of Disturbing Loads. The advent of electronic power conversion has been widely applauded by users, but the drawbacks from the point of view of power quality have not always been recognized. The very advantages of solid-state devices that made possible modern switching power supplies, inverter-rectifiers, high-frequency induction heating, and adjustable-speed drives also make these power converters into generators of harmonic currents and additional sources of line voltage drops. Thus, in addition to the disturbances generated by the normal operation of the familiar power delivery and load equipment, the disturbances resulting from the new electronic loads must be taken into consideration.

Harmonic currents caused by many types of customer load and utility equipment provide an example of this complexity. For many years, harmonic currents originated mainly from a few major sources, such as arc furnaces and high-voltage dc transmission terminals. In these cases, they could be removed with relative ease by placing a large (and expensive) filter between the source and the main power line. Today, however, significant power line harmonics are being caused by many small, widely dispersed customer loads, such as rectifiers and solid-state controls for adjustable-speed motors. At the same time, an increasing number of other customers are using sensitive equipment, such as computers, the operation of which may be adversely affected by harmonics.

It would not be economically feasible to detect and filter each small source of harmonics or to isolate each sensitive load from all power line disturbances. A more practical approach is to control harmonics by agreement on limits for emission levels with filters installed on major offending loads, while defining acceptable susceptibility level for equipment. Unusually sensitive electronic equipment may be supplied by special power-conditioning interfaces, external to or incorporated with the equipment. Such an approach will require collaboration among utilities, equipment manufacturers, users, regulatory agencies, and standards-setting bodies.

3.1.3 The Concept of Power Quality. Emerging concerns over these issues has resulted in focusing attention on the quality of the power necessary for successful operation of diverse loads, on practical limits to the capability of delivering power of high quality to diverse customers, and on the economics of the producer-user partnership. The term "power quality" is now widely used, but objective criteria for measuring the quality of power—a prerequisite for quantifying this quality—need better definition. A high level of power quality is understood as low level of disturbances; agreement on acceptable levels of disturbances is needed. As one step in that direction, 3.4 presents an overview of the nature of disturbances.

Another difficulty in assessing the need for an interface between the utility power and sensitive loads is the subjective nature of estimating the cost of equipment misoperation attributable to power disturbances. This particular aspect is addressed from the technical point of view in this book, but the detailed economics are beyond its scope. As generic information, Chapter 8 provides guidance on the selection of line-conditioning equipment.

3.1.4 Conflicting Design Philosophies and Safety. The issue of power quality is made more confusing by conflicting philosophies advocated by people of different technical backgrounds and commercial interests. An example of this problem is found in the apparent conflicts resulting from interpretations of grounding requirements. The general requirement of a safe configuration and a safe operation for a power system is endorsed by all parties (utilities, users, regulatory bodies, voluntary standards organizations, etc.), but in some instances these requirements translate into wiring practices alleged to interfere with smooth operation of electronic systems. Many anecdotal case histories have been encountered where system designers complain that the requirements of the National Electrical Code (NEC) (ANSI/NFPA 70-1993 [1][6]) prevent their system from operating in a satisfactory (safe) manner. This apparent conflict of philosophy can only be settled by giving safety the prevailing directive. That prevailing directive is precisely the purpose of the NEC, and correct application of the NEC directives does *not* prevent satisfactory operation of properly wired and grounded installations. If any adaptations have to be made for the system to operate satisfactorily, the equipment manufacturer must incorporate them in the equipment design, rather than ask for deviations from the NEC.

3.2 Power Quality Considerations

3.2.1 General Discussion. Power systems operate with a constant line voltage, supplying power to a wide variety of load equipment. Power levels range from a few watts to megawatts, and the voltages at which the energy is generated, transported, and distributed range from hundreds of volts to hundreds of kilovolts. Transmission and primary distribution of this power are made at high voltages, tens to hundreds of kilovolts, in order to provide efficient and economic transportation of the energy over long distances. Final utilization is generally in the range of 120 V (typical residential) to less than one thousand (industrial), and a few thousands for large loads.

At all these voltage and power levels, no matter how high, the equipment is dependent upon maintenance of a normal operating voltage because it has only limited capability of withstanding voltages exceeding the normal level. At lower than normal levels, the equipment performance is generally unsatisfactory, or there is a risk of equipment damage. These two disturbances, excessive voltage and insufficient voltage, are described with different names

---

[6] The numbers in brackets correspond to those of the references in 3.9; when preceded by the letter "B," they correspond to those in the bibliography in 3.10.

depending on their duration. There are also types of disturbances, as described in 3.2.2, involving waveform distortion and other deviations from the expected sine wave.

3.2.2 Classification of Disturbances. Four power system parameters—frequency, amplitude, waveform, and symmetry—can serve as frames of reference to classify the disturbances according to their impact on the quality of the available power. A brief discussion is given below of the need for evaluation of their impact on sensitive loads.

Frequency variations are rare on utility connected systems, but engine-generator-based distribution systems can experience frequency variations due to load variations and equipment malfunctions.

Amplitude variations can occur in several forms; their description is inextricably associated with their duration. They range from extremely brief durations to steady-state conditions, making the description and definition difficult, even controversial at times. Their causes and effects need close examination to understand the mechanisms and to define an appropriate solution.

Waveform variations occur when nonlinear loads draw a current that is not sinusoidal. One could also describe an amplitude variation as momentary waveform variation, but the intended meaning of the term is a steady variation of the waveform, or lasting at least over several cycles. This type of disturbance may be described as harmonic distortion because it is easy to analyze as the superposition of harmonics to the nominal frequency of the power system.

Dissymmetry, also called unbalance, occurs when unequal single-phase loads are connected to a three-phase system and cause a loss of symmetry. This type of disturbance primarily concerns rotating machines and three-phase rectifiers and, as such, is not receiving broad attention. It is important, however, for machine designers and users. The percentage by which one-phase voltage differs from the average of all three is the usual description of this type of disturbance.

3.2.3 Origin of Disturbances. The term "origin of disturbances" can be understood in two different manners. According to one interpretation, the concern is over the source of the disturbance and whether it is external or internal to the particular power system. Typically, the boundary of a power system is defined as the watthour meter, and reference is made to the "utility side of the meter" (external source), or to the "user side of the meter" (internal source).

According to another interpretation, the concern is over the nature of the source; the disturbance is then described in technical terms, such as lightning, load switching, power system fault, and nonlinear loads. Depending on local conditions, one can be more important than the others, but all need to be recognized. The mechanism involved in generating the disturbance also determines whether the occurrence will be random or permanent, unpredictable or easy to define.

The first interpretation is motivated by the goal of assigning responsibility for the problem, and possibly liability for a remedy. The second interpretation is motivated by the goal of understanding the problem and developing a technically sound remedy. When discussing the problem among the parties involved, the different points of view must be recognized, lest misunderstandings occur. In the following paragraphs, the second interpretation leads to a description of mechanisms producing the disturbances.

The general tendency of users is to attribute most of their disturbance problems to the utility source. Many other sources of disturbances, however, are located within the building and are attributable to operation by the end-user of other equipment. Finally, there are sources of disturbances—or system errors—not associated with the power input of the equipment, such as electrostatic discharges to the equipment enclosure or cables, radiated electromagnetic interference, ground potential differences, and operator errors (Fig 3-1).

Some disturbances occur at random and are not predictable for a given site and are not repeatable although statistical information may be available on their occurrence (IEEE Std C62.41-1991) [7], IEEE Std 519-1992 [12]). Other disturbances, especially those associated with operation of other equipment, can be predicted, are repeatable, and can be observed by performing the operating cycle of that equipment.

Lightning surges are the result of direct strikes to the power system conductors as well as the result of indirect effects. Direct strikes inject the total lightning current into the system. The current amplitudes range from a few thousand amperes to a few hundred thousand amperes. However, the rapid change of current through the impedance of the conductors produces a high voltage that causes secondary flashover to ground, diverting some current even in the absence of an intentional diverter. As a result, equipment connected at the end of overhead conductors are rarely exposed to the full lightning discharge current. Indirect effects include induction of overvoltages in loops formed by conductors and ground potential rises resulting from lightning current in grounding grids or the earth.

A lightning strike to the power system can activate a surge arrester, producing a severe reduction or a complete loss of the power system voltage for one half-cycle. A flashover of line insulators can trip a breaker, with reclosing delayed by several cycles, causing a momentary power outage. Thus, lightning can be the obvious cause of overvoltages near its point of impact, but also a less obvious cause of voltage loss at a considerable distance from its point of impact. Clearly, the occurrence of this type of disturbance is unpredictable at the microscopic level (e.g., specific site). At the macroscopic level (e.g., general area), it is related to geography, seasons, and local system configuration.

Induction of surges by nearby lightning discharges is a less dramatic but more frequent event. The resulting surge characteristics are influenced not only by the driving force—the electromagnetic field—but also by the response of the power system—its natural oscillations. This dual origin makes a general description of the occurrence impractical, but nevertheless a consensus exists on representative threats for various environments.

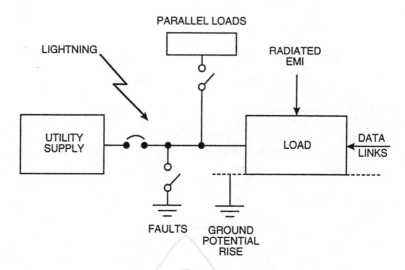

(a) Examples of Interference That Do Not Involve Utility Supply

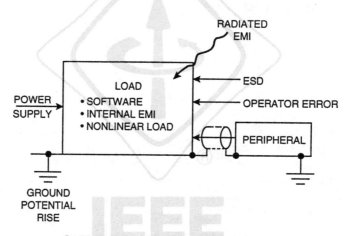

(b) Example of Source Interference

Fig 3-1
Sources of Disturbances

Load switching is a major cause of disturbances. Whenever a circuit containing capacitance and inductance is being switched on or off, a transient disturbance occurs because the currents and voltages do not reach their final value instantaneously. This type of disturbance is inescapable and its severity depends on the relative power level of the load being switched and of the short-circuit current of the power system in which the switching takes place. Switching large loads on or off can produce long-duration voltage changes beyond the immediate transient response of the circuit. Whether the switching is done by the utility or by the user is immaterial from a technical point of view, although the responsibility may be the subject of a contractual dispute. More complex circuit phenomena, such as current chopping, prestrikes, and restrikes, can produce surge voltages reaching ten times the normal circuit voltages, involving energy levels determined by the power rating of the elements being switched. These complex surges can have very destructive effects, even on rugged equipment, and must be controlled at the source as well as mitigated at the loads.

The occurrence of load switching disturbances is somewhat predictable, but not necessarily under controlled conditions. The introduction of power conversion equipment and voltage regulators operating by switching on and off at high frequency has created a new type of load switching disturbance. These disturbances occur steadily, although their amplitude and harmonic content will vary for a given regulator as the load conditions vary.

Power system faults occur on both sides of the meter, resulting from equipment failure or external causes (vehicle collisions, storms, human errors). These disturbances can range from a momentary voltage reduction to a complete loss of power lasting for minutes, hours, or days. Their accidental origin makes them unpredictable, although the configuration of a power system and its environment can make it more or less prone to this type of disturbance [B7].

Nonlinear loads draw nonsinusoidal currents from the power system, even if the power system voltage is a perfect sine wave. These currents produce nonsinusoidal voltage drops through the system source impedance which distort the sine wave produced by the power plant generator. A typical nonlinear load is a dc power supply with capacitor-input filter, such as used in most computers, drawing current only at the peaks of the voltage sine wave. This situation has also created a new concern, that of insufficient ampacity of the neutral conductor in a three-phase system feeding power supplies; see Chapter 4 for a discussion of this problem.

Other disturbances. The discharge of electrostatic charges built upon the human body or objects, can also inject unwanted voltages or currents into the circuits. This phenomenon is associated with operator contact with the equipment (keyboards, panel switches, connectors) rather than with the quality of the incoming power. Thus, it is not included in the scope, but should of course not be ignored when troubleshooting equipment problems.

Another concern is related to nuclear explosions effects, under the acronym of NEMP (nuclear electromagnetic pulse). However, concerns over this type of "disturbance" are beyond the scope of this standard, and are the subject of specialized documents.

3.2.4 Power Quality Site Surveys. Power quality site surveys have been performed and reported and by a number of investigators. However, the reports are difficult to compare because the names of the disturbances and their thresholds vary among the reports. Manufacturers of disturbance recorders have defined the events reported by their instruments at some variance with other sources of definitions. To help resolve the confusion, a new IEEE Working Group has been chartered and is attempting to provide unique definitions for the disturbances. The results of this effort, however, will take some time to be generally recognized and accepted. In the meantime, terms used by different authors might have different meanings, leaving on authors the burden of defining their terms and leaving on readers the burden of being alert for possible ambiguities.

One example of such ambiguities occurs when attempting to summarize data from different surveys. For instance, two surveys have been widely cited ([B2], [B5]); each was aimed at defining the quality of power available for the equipment of concern to the authors. As a result, each author categorized the disturbances according to the criteria significant to that equipment, including the threshold below which disturbances are not recorded by the instrument. With hindsight, it is not surprising that the criteria were different; when comparing the data from the two surveys expressed in percentages (leading to pie chart representations by some authors of application papers), a puzzling difference was found. By analyzing the detail of the survey premises and definitions, the differences can be reconciled to some extent [B9].

3.3 Grounding Considerations. Proper grounding is essential to safe and satisfactory performance of a power system. There are three requirements for such grounding:

(1) Providing a low-impedance path for the return of fault currents, so that an overcurrent protection device can act quickly to clear the circuit,
(2) Maintaining a low potential difference between exposed metal parts to avoid personnel hazards, and
(3) Overvoltage control.

A very comprehensive discussion of these considerations, applicable to any installation, can be found in other books in the IEEE Color Book series: IEEE Std 141-1986 [8], IEEE Std 142-1991 [9], and IEEE Std 446-1987 [10].

Grounding of sensitive equipment, such as information technology equipment, involves another aspect because this type of equipment, by design, communicates with other equipment by data lines. These data lines often carry their own zero-reference conductors that may or may not be bonded to the equipment safety ground. Thus, there may be a common path between the signal circuits and the power circuits, giving rise to problems of noise coupling. These data lines generally carry high-frequency signals, so that impedance considerations related to the power-frequency safety aspects of a grounding system may not necessarily provide the desirable low impedance at the signal frequencies. Worst yet, conflicts may arise between the perceptions of signal engineers on the goals of their grounding practices and the non-negotiable safety requirements of power system grounding practices.

In the case of computer systems distributed throughout a facility, the situation is different. The local ground at the point where a terminal is connected to the power system is likely to be at a high-frequency potential different from those of the other elements of the system. This concern is briefly addressed in 3.7 of this chapter. A comprehensive discussion of the high-frequency aspects of grounding practices for mainframe computer rooms can be found in [B4].

GENERAL NEEDS GUIDELINES

3.4 Protection Against Disturbances

3.4.1 General. The concept of protection implies the confrontation of a hostile environment and susceptible equipment. Protection of the equipment against the hostile environment is the goal of the technology of electromagnetic compatibility. Discussing the need for protection, therefore, takes on two aspects: characterizing the environment, and characterizing the susceptibility of the equipment. Disturbances to the environment have been briefly discussed in the preceding paragraphs. More complete descriptions can be found in other IEEE standards (IEEE Std 519-1992) [12], (IEEE Std C62.41-1991) [7].

The susceptibility level of the equipment, however, is a subject that is more difficult to quantify because it requires the disclosure by manufacturers of information that some are reluctant to provide, lest it be misunderstood or misused. Nevertheless, the consensus process has produced a useful graph of typical susceptibility levels—or the converse, tolerance levels. This graph has been widely published, and is reproduced here as Fig 3-2. Note that the graph only addresses the magnitude of the voltage, with a corresponding duration

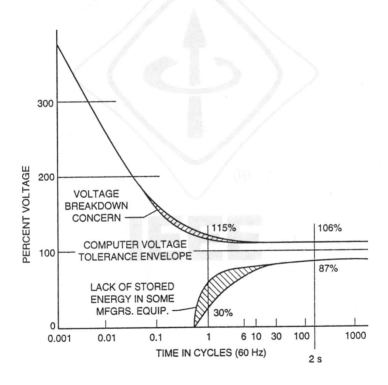

Fig 3-2
Typical Design Goals of Power-Conscious Computer Manufacturers

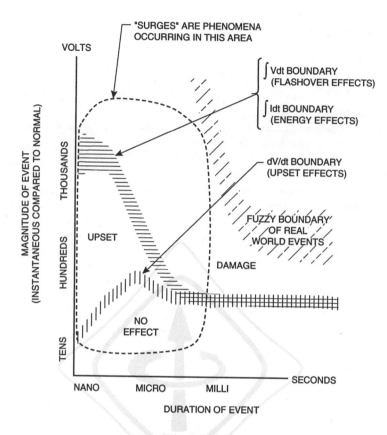

**Fig 3-3
Relationship Between Amplitude, Duration, Rate of Change of Disturbances,
and Their Effects on Equipment**

of the disturbance. One aspect that it does not address is the significance of the rate of change in voltage disturbances. This rate of change is important in two aspects:

(1) A fast rate of change has greater capability of producing a disturbance in adjacent circuits by capacitive and inductive coupling, and
(2) A slow rate of change can make ineffective a protective device based on inserting a series inductance in the power line.

Detailed analysis of the rate of change issue is beyond the scope of this chapter, but Fig 3-3, inspired from Fig 3-2, takes the concept one step further in identifying the issues.

3.4.2 Noise Protection. Noise on the power line is generally understood as a disturbance of low amplitude, a small fraction of the system voltage, while a surge on the power line is generally understood as a disturbance of larger fraction, or multiple of the system voltage. The boundary between the two phenomena is not clear, and documents prepared by groups of different backgrounds and interest can vary on the definition of this boundary. Noise effects are often lumped under the label of electromagnetic interference (EMI) and addressed by frequency-domain-oriented specialists. Surge effects are generally addressed by time-domain-oriented specialists more concerned with damaging effects than upset effects. These different points of view are also reflected in Fig 3-3. IEEE Std 518-1982 [11] and textbooks [B12] and [B13] provide a comprehensive discussion of noise-reduction practices.

3.4.3 Surge Protection. Surges can have many effects on equipment, ranging from no detectable effect to complete destruction. In general, electromechanical devices withstand voltage surges until a dielectric breakdown occurs, while electronic devices can have their operation upset before hard failure occurs. At intermediate levels, progressively more intense upset occurs, until breakdown takes place. Definitions of the level beyond which a transient overvoltage becomes a threat depend on the type of victim equipment. While electromechanical devices can generally tolerate voltages of several times their rating for short durations, few solid-state devices can tolerate much more than twice their normal rating. Furthermore, data processing equipment can be affected by fast changes in voltage with relatively small amplitude compared to the hardware-damaging overvoltages.

Survival or undisturbed operation of the equipment can be achieved in three manners: eradication of the cause of surges (for instance, the elimination of lightning . . .), building equipment immune to any level of surges, no matter how high, or, the obvious choice, finding the best economic trade-off. Moderate surge-withstand capability is built into equipment, and the worst surges occurring in the environment are reduced, by application of suitable protective devices, to a level that the equipment can tolerate [B1].

Low-voltage, end-user type surge protective devices are often described as "transient suppressors," but their operation is really a diversion of the surge current through a low-impedance path preventing the rise of high voltages across the load terminals. For large surge currents, this diversion is best accomplished in two stages: The first diversion should be performed at the entrance to the building, typically by conventional surge arresters rated for this duty. Then, any residual voltage resulting from the action of the arrester can be dealt with by a second protective device at the power panel of the computer room or at the terminals of a connected load. In this manner, the wiring inside the building is not required to carry the large surge current to and from the diverter at the end of a branch circuit. Such a long path for the current would produce inductive voltage drops in the branch circuit wires, resulting in a rise of the neutral or grounding conductors terminals with respect to local grounds. A potential problem, however, is associated with the two-stage protection scheme: if not properly coordinated, the second protective device may attempt to divert all of the impinging surge and fail in the process. Thus, proper attention must be given to coordination of cascaded surge protective devices [B8].

3.4.4 Sag Protection. Sag protection is not addressed here. See 8.2.8 and 8.2.9, and IEEE Std 446-1987 [10] for more information.

3.5 Safety Systems. Safety systems protect life and property from damage or loss due to accidents. For equipment, the degree of protection should be based on the value and criticality of the facility. Personnel safety, of course, is not dependent on these criteria. Defining this degree requires an in-depth knowledge of the installation and its function. The following questions should be considered when designing these systems:
  (1) How long will it take to replace the equipment and at what cost?
  (2) Can the function of the facility be performed elsewhere?
  (3) Is the loss that of a key component, resulting in operation interruption?

Safety systems can be as simple as a manually operated Emergency Power Off button, or as complex as a fully interlocked system. However, the more complex a fully integrated system becomes, the higher the probability of system confusion or failure. Typical systems include the following functions:
  (1) Smoke and fire protection
  (2) Environmental control
  (3) Smoke exhaust
  (4) Fire extinguishing
  (5) Emergency lighting
  (6) Security

The interfacing of a safety system is generally unique for each installation and requires a logical design approach. Through a well-defined logic matrix and sequence priorities, it is possible to develop a system that can be maintained, modified, or expanded with little confusion and minimum expense.

Generally, safety systems operate from 120 V ac, 24 V ac, or 24 V and 12 V dc. In any case, these systems must remain powered at all times. The quality of the power supplied to these systems is as important as that of the power delivered to the information technology system. Disturbances in the power supply of the safety system can cause shutdown of the protected system.

3.6 Information Technology Systems. Under the name of information technology systems we classify all equipment incorporating at least two ports: a data port for input and output of signals, and a power supply port. The data port can be linked to the public telephone network, to a dedicated terminal, or to a communications bus or system. The significant aspect of such equipment is its two-port configuration; in many instances, the power port design and connections are regulated by one set of standards, while the data port is regulated by another set of standards, if any.

Safety aspects of grounding practices are fulfilled with no conflict by power system designers. On the other hand, designers of information technology systems may have different criteria or practices from those of the power system designers. Signal circuits are not always grounded by a low (zero) impedance bond to their equipment (chassis, enclosure) ground. Some of these systems use a reference that is grounded. Others use balanced pairs that may or may not carry their own ground reference. However, at the high frequencies associated with disturbances, all circuits are capacitively coupled to ground, and inductively coupled to adjacent circuits. Therefore, noise can be injected in these data circuits by power system ground or fault currents, by electromagnetic interference from other systems or lightning, and by other sources. Remedies to noise problems proposed by information technology specialists are sometimes at variance with the requirements for effective grounding from the point of view of power system faults or lightning current protection.

One especially troublesome problem is that of systems featuring several elements in different locations, therefore powered from different branch circuits, but linked by a data cable that carries it own zero reference—a conductor linking the grounding connections in the different locations. Under moderate conditions, the ground loop thus formed can couple noise into the signal path. Under more severe conditions, such as a power system fault or a surge being diverted through the grounding conductors, substantial differences can exist between the "ground" potential of two distant elements of the system; this difference of potential can cause component failures in the circuits.

3.7 Shielded, Enclosed EMI/EMC Areas (TEMPEST)

3.7.1 General. Electromagnetic interference and electromagnetic compatibility (EMI/EMC) requirements are intended to limit the spurious emissions given off by electronic equipment and to ensure that electronic equipment is not adversely affected by such emissions. Typical EMI/EMC requirements are contained in [B3] or in documents promulgated by Technical Committee 77 (Electromagetic Compatibility) of the International Electrotechnical Commission (IEC). The requirements implied by "TEMPEST" have different motivations. TEMPEST is a government term referring to the concerns over compromising emanations from any information processing equipment. Many years ago, Department of Defense personnel learned that it is possible to intercept the radio emissions given off by electronic equipment and that, with the aid of computers, classified information could be extracted from these signals by unauthorized parties. As the use of computers has become more commonplace in the office and the "decoding" business, the probability of such interceptions has increased.

TEMPEST requirements are usually achieved by placing a shielded enclosure around the equipment emanating the compromising signal. EMC requirements are achieved the same way. This metal enclosure reflects or absorbs the signals and attenuates them to an undetectable level. In recent years, TEMPEST interest has increased in nongovernment agencies. Some computer manufacturers now offer TEMPEST shielded computers and peripherals for commercial use.

3.7.2 Electrical Safety Requirements. Shielding hardware and power distribution system designed to meet the objectives of EMI/EMC and TEMPEST must always meet the requirements of the NEC [1]). In particular, the grounding and bonding of shields and associated components must comply with Article 250 of the NEC. Distribution systems and equipment within the shielded area are bonded to the interior of the shield while the outside of the shield is bonded to the facility grounding system [B10]. Although this external connection has little or no effect on the equipment within the shield, it is essential to prevent the enclosure from reaching dangerous potentials relative to its surroundings.

3.7.3 Basic Requirements. A Faraday cage that provides an electromagnetic and radio frequency shield enveloping the equipment to be protected best describes the basic requirements of EMI/EMC and TEMPEST. This shield isolates the protected circuits from spurious external signals and also attenuates TEMPEST emanations to levels that are too small to be intercepted or analyzed. To be usable, this shield must have penetrations for personnel and equipment access, power lines, control cables, and ventilation. The number of shield penetrations must be held to a minimum since each penetration is a potential leakage source and will require additional maintenance. For those penetrations that cannot be eliminated, proper construction to eliminate leaks is essential. Also, equipment and hardware installed within the shielded area must comply with EMI/EMC requirements in order to tolerate any residual internal electromagnetic fields. Topological grounding methods should also be employed. That is, each shielded region (topological zone) should have a separate grounding system making contact with both the inner and outer shield defining the zone [B6]. For more information on shielded areas, see [B10] and [B11].

3.8 Coordination with Other Codes, Standards, and Agencies

3.8.1 General. There is a large body of guidelines, standards, and codes that address the issues of power quality, safety, and operational integrity of a power system and its connected equipment. These documents are prepared by diverse organizations, including voluntary consensus standards such as the IEEE documents, national position standards such as the recommendations of the IEC, safety standards such as those of the Underwriters Laboratories (UL), performance standards prepared by users' organizations, interchangeability standards prepared by manufacturers trade organizations, and regulatory standards promulgated by local and national agencies.

While conflicts are not intended among these documents, the wide diversity of needs and points of view unavoidably creates ambiguities at best and conflicts at worst. As indicated earlier, however, the safety and legal aspects of any conflict mandate a prevailing role for the NEC [1].

3.8.2 National Electrical Code (NEC). The NEC [1] is a document prepared by consensus of a number of panels where national experts develop a set of specific and detailed requirements. These requirements are based on long-established practices, complemented by a permanent review process with a three-year cycle. The NEC is generally adopted by local jurisdictions, either in its entirety or with some modifications, and thus becomes enforceable by local inspection authorities. Conspicuous exceptions exist, however, in the domain of application: the power generation and distribution facilities of electric utilities are not regulated by the NEC, but have their own safety standards; United States government facilities are not regulated by the NEC, although installations are generally made in accordance with the NEC. Some jurisdictions, notably large cities in the United States, have their own local codes usually based on the NEC with additional requirements.

3.8.3 Underwriters Laboratories (UL) Standards. UL is an independent, not-for-profit organization operating in the field of public safety. It operates product safety certification programs to determine that manufactured materials and products produced under these programs are reasonably safeguarded against foreseeable hazards. UL publishes standards, product directories, and other information. Approximately 500 published standards now exist. These standards are generally recognized by inspection authorities in the United States. Note, however, that there are other competent testing agencies that can conduct certification programs based upon UL Standards.

3.8.4 Other Laboratories and Testing Agencies. Other laboratories and testing agencies have also performed tests on equipment, for the purpose of listing or for providing an independent verification of performances. The Occupational Safety and Health Administration (OSHA) requires listing only by a "recognized" testing agency, without defining such agencies.

3.8.5 National Electrical Manufacturers Association (NEMA) Standards. NEMA develops product standards, some of which are recognized as Accredited Standards Committee standards. These standards are generally concerned with equipment interchangeability, but also contain documentation on operation and safety features.

3.8.6 National Institute of Standards and Technology (NIST). NIST (formerly the National Bureau of Standards) is a United States Government agency established initially for the purpose of maintaining standards of measurements and calibration of instruments, including traceability. Over the years, the role of NIST has expanded to include a broad range of research activities. The staff of NIST is active in many standard-writing groups, through individual contributions of experts in each specific field. However, NIST does not promulgate standards in the meaning of documents such as IEEE, IEC, or ANSI standards. A notable exception is a developing series of standards under the category of Federal Information Processing Standards Publications, such as [B4].

3.8.7 International Standards. International standards are developed by a different process than the typical voluntary standard process used in the United States, as exemplified by the present book. The prevalent set of standards is developed by the IEC, and covers most of the engineering and application aspects of electromechanical and electronic equipment. Technical Committees involved in the development of documents related to power and grounding include the following:

- *Technical Subcommittee 28A*, for insulation coordination concerns. A report prepared by this subcommittee (IEC Publication 664-1980 [2]) discusses in detail an approach whereby overvoltage categories would be assigned to various types of equipment. The overvoltage capability of the equipment would become part of the equipment nameplate information, ensuring proper installation in known environments.
- *Technical Committee 64,* for fixed (premises) wiring considerations.
- *Technical Committee 65 WG4,* for electromagnetic compatibility of industrial process control equipment. This working group has produced and continues to update a family of documents addressing surge immunity, fast transients, and electrostatic discharges (IEC Publications 801-1 (1984) [3], 801-2 (1984) [4], 801-3 (1984) [5], 801-4 (1988) [6]).
- *Technical Committee 77,* for electromagnetic compatibility. Within the broad scope of all possible disturbances to electromagnetic compatibility, this committee is developing documents related to conducted disturbances. These documents are generic descriptions and classifications of the environment, leading to the specification of immunity tests in general. Detailed test specifications for a given equipment are left to the relevant product committee.

3.9 References. This standard shall be used in conjunction with the following publications. When the following standards are superseded by an approved revision, the revision shall apply:

[1] ANSI/NFPA 70-1993, National Electrical Code.[7]

[2] IEC Publication 664-1980, Insulation co-ordination within low-voltage systems including clearances and creepage distances for equipment.[8]

[3] IEC Publication 801-1 (1984), Electromagnetic compatibility for industrial-process measurement and control equipment, Part 1: General introduction.

[4] IEC Publication 801-2 (1984), Electromagnetic compatibility for industrial-process measurement and control equipment, Part 2: Electrostatic discharge requirements.

---

[7] NFPA publications are available from Publications Sales, National Fire Protection Association, 1 Batterymarch Park, P.O. Box 9101, Quincy, MA 02269-9101, USA.

[8] IEC publications are available from IEC Sales Department, Case Postale 131, 3 rue de Varembé, CH 1211, Genève 20, Switzerland/Suisse. IEC publications are also available in the United States from the Sales Department, American National Standards Institute, 11 West 42nd Street, 13th Floor, New York, NY 10036, USA.

[5] IEC Publication 801-3 (1984), Electromagnetic compatibility for industrial-process measurement and control equipment, Part 3: Radiated electomagnetic field requirements.

[6] IEC Publication 801-4 (1988), Electromagnetic compatibility for industrial-process measurement and control equipment, Part 4: Electrical fast transient/burst requirements.

[7] IEEE Std C62.41-1991, IEEE Recommended Practice on Surge Voltages in Low-Voltage AC Power Circuits (ANSI).[9]

[8] IEEE Std 141-1986, IEEE Recommended Practice for Electrical Power Distribution for Industrial Plants (IEEE Red Book) (ANSI).

[9] IEEE Std 142-1991, IEEE Recommended Practice for Grounding of Industrial and Commercial Power Systems (IEEE Green Book) (ANSI).

[10] IEEE Std 446-1987, IEEE Recommended Practice for Emergency and Standby Power Systems for Industrial and Commercial Applications (IEEE Orange Book) (ANSI).

[11] IEEE Std 518-1982 (Reaff 1980), IEEE Guide for the Installation of Electrical Equipment to Minimize Electrical Noise Inputs to Controllers from External Sources (ANSI).

[12] IEEE Std 519-1992, IEEE Guide for Harmonic Control and Reactive Compensation of Static Power Converters (ANSI).

3.10 Bibliography

[B1] IEEE Std C62-1990, Complete 1990 Edition: Guides and Standards for Surge Protection.

[B2] Allen, G. W. and D. Segall, "Monitoring of Computer Installations for Power Line Disturbances," *IEEE Winter Power Meeting Conference Paper*, WINPWR C74 199-6, 1974 (abstract in *IEEE Transactions on PAS*, Vol. PAS-93, Jul/Aug 1974, p. 1023).

[B3] CFR (Code of Federal Regulations), Title 47, Part 15, published by the Office of the Federal Register.

[B4] Federal Information Processing Standards Publication 94: *Guideline on Electrical Power for ADP Installations*, National Technical Information Service, 1983.

---

[9] IEEE publications are available from the Institute of Electrical and Electronics Engineers, Service Center, 445 Hoes Lane, P.O. Box 1331, Piscataway, NJ 08855-1331, USA.

[B5] Goldstein, M. and P. D. Speranza, "The Quality of U. S. Commercial ac Power," *INTELEC* (IEEE International Telecommunications Energy Conference), 1982, pp. 28–33 [CH1818-4].

[B6] Graf, W. and J. E. Nanevicz, "Topological Grounding Anomalies," International Aerospace and Ground Conference on Lightning and Static Electricity, June 20–28, 1984.

[B7] Key, T. S. "Diagnosing Power Quality Related Computer Problems," *IEEE Transactions on Industry Applications*, Vol. IA-15, No. 4, July/Aug 1979.

[B8] Martzloff, F. D. "Coordination of Surge Protectors in Low-Voltage AC Power Circuits," *IEEE Transactions on Power Apparatus and Systems*, Vol. 99, No. 1, Jan/Feb 1980, pp. 129–33.

[B9] Martzloff, F. D. and T. M. Gruzs, "Power Quality Surveys: Facts, Fictions, and Fallacies," *IEEE Transactions on Industry Applications*, Vol. 24, No. 6, Nov/Dec 1988, pp. 1005–18.

[B10] MIL-HNDBK-419, Grounding, Bonding, and Shielding for Electronic Equipments and Facilities, Vol. 1 (Basic Theory), Vol. 2 (Applications).

[B11] MIL-STD-188/124, Grounding, Bonding and Shielding for Common Long Haul Tactical Communication Systems Including Ground Based Communications—Electronics Facilities and Equipments.

[B12] Morrison, R. *Grounding and Shielding in Instrumentation*, New York: John Wiley & Sons, 1977.

[B13] Ott, H. *Noise Reduction Techniques in Electronic Systems*, New York: John Wiley & Sons, 1989.

Chapter 4
Fundamentals

4.1 Introduction. Successful, reliable operation of sensitive electronic equipment requires adherence to the fundamentals of physics. This chapter reviews appropriate fundamental concepts, with the objective of establishing an appreciation of how things work and their related failure modes. This focus on fundamentals prepares the reader for recommended design practices given in Chapter 9.

Rapid changes in the electronics and communications industries make it almost impossible for design engineers to be expert in all related disciplines. Therefore, a further objective of this chapter is to forge a consensus on related design issues and the expression of these issues via a common language.

4.2 Impedance Considerations. An understanding of electrical impedance is fundamental to the design of power systems for sensitive electronics. The total system impedance can be grouped into three fundamental parts: the power source, the distribution, and the load impedances. It is important to note that the nature and magnitude of these impedances vary with frequency. They and their frequency related considerations are discussed in this section.

4.2.1 Frequencies of Interest. The most distinguishing characteristic of power systems, for sensitive electronic equipment, is that they must behave in an orderly fashion from dc to tens of megahertz. This total frequency range can be conceptualized as two distinct frequency ranges: a power/safety range and a performance range.

4.2.1.1 Power/Safety Range. The power/safety range typically encompasses a frequency range from dc to several tens of harmonics above the power source's nominal frequency (e.g., 60 Hz). Impedances in this range tend to be dominated by lumped resistance, inductance, and capacitance. Designers of typical industrial and commercial power systems are generally familiar with the needs and design standards of this frequency range (National Electrical Code (NEC) (ANSI/NFPA 70-1993 [2]),[10] IEEE Std 446-1987 [10]).

4.2.1.2 Performance Range. The performance range extends from dc to tens of megahertz. The upper portion of this range has historically been the domain of radio frequency engineers, and in general is identified as a specialty area, distinctly different from power engineering. The term

---

[10] The numbers in brackets correspond to those of the references in 4.9; when preceded by the letter "B," they correspond to those in the bibliography in 4.10.

performance range is defined here to denote that the conducted and radiated electromagnetic energy (in the frequency range between tens of kilohertz and tens of megahertz) can significantly impact the operational performance of sensitive electronic equipment. Impedances in this range tend to be characterized by distributed resistive, inductive, and capacitive elements, particularly at the higher frequencies (FIPS PUB 94 [11]).

4.2.2 Power Source Impedance. Knowledge of the power source impedance is critical to the understanding of critical load-source interactions. Power source impedance is the ratio of incremental internal voltage drop within the same source, dE, to the incremental load current supplied by that source, dI. That is, the impedance is $Z = dE/dI$. Impedance of a power source can be further delineated as internal, forward-transfer, and output impedance. These basic concepts of source impedance can be illustrated by a simplified equivalent diagram of a transformer. Fig 4-1 shows such a diagram where the magnetizing inductance of the core is neglected.

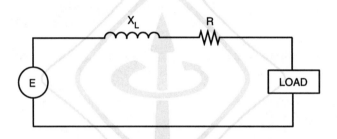

Fig 4-1
First Order Model of Transformer Impedances

4.2.2.1 Internal Impedance. Internal impedance is the impedance of the power source at its design frequency. For example, the determination of a transformer percent internal impedance (%Z) requires a knowledge of the following:
 (1) The input voltage necessary to make the current in a short-circuited secondary equal to the rated current.
 (2) The rated input voltage.

The transformer internal impedance, expressed as a percent, is the ratio of (1) above divided by (2), then multiplied by 100. The internal impedance is often provided on transformer nameplates.

A transformer with a 5% internal impedance allows 20 times its rated current to flow during short-circuit conditions, assuming sufficient fault current is available on its primary. Typical power transformers suitable for electronic equipment are identified in [B25]. These transformers tend to have impedances in the range of 3–6% at their nominal design frequency (e.g., 60 Hz).

FUNDAMENTALS

It is desirable to have a low-internal impedance, such that supply voltage variances are small for normal swings in load currents. If the source impedance is too low, possible short-circuit current can be excessive to the point that special circuit breakers or supplementary current limiting fuses are required to interrupt fault current. A suitable internal impedance range for transformers, which serve nonlinear load currents for sensitive electronic equipment, is 3–6%.

4.2.2.2 Forward Transfer Impedance. Forward transfer impedance is an attribute similar to internal impedance, but at frequencies other than the nominal power system frequency (e.g., 60 Hz). Knowledge of the forward transfer impedance allows the designer to assess the capability of the power source to provide load current at the harmonic frequencies needed to preserve a suitable output voltage waveform. Generally, the highest frequency of interest is 3 kHz for 50–60 Hz power systems, and 20–25 kHz for nominal 400 Hz power systems (which is about 50 times the supply frequency).

A common method for determining forward transfer impedance of transformers and filters is to measure simultaneously the input signal voltage and short-circuited output current. The ratio is the forward transfer impedance.

Generally, the forward transfer impedance will increase with increasing frequency. It is desirable to have a low forward transfer impedance at the nominal power frequency and its low-order harmonics (e.g., up to 50th harmonic). At frequencies above the 50th harmonic, a high value of forward transfer impedance is desirable to attenuate transient voltages conducted by the power system.

4.2.2.3 Output Impedance. Output (reverse transfer) impedance of a power source is an attribute similar to forward transfer impedance but it describes the impedance of the power system as seen from the load.

If the load generates harmonic currents, then these currents flow back towards the source of power. These currents produce voltage drops across the source impedance which add to (or subtract from) the power system voltage. The amplitude and waveshape of the line voltage can change significantly. Therefore, it is very important that the power source have low-output impedance to these harmonic currents. At higher frequencies than those produced by harmonics, a high-output impedance provides some filtering of high-frequency transients before these reach the load. Output impedances generally rise with frequency but distributed capacitances can allow resonance that may lower output impedance at specific frequencies.

Distributed primary to secondary capacitance in transformers can act to accentuate higher frequency transients on the load side relative to the power frequency. Adding shields within the transformer can reduce the primary to secondary capacitance. Using additional capacitors and filters to attenuate these higher frequency transients can have unexpected and undesirable consequences if shunt capacitors (which are usually involved) elevate the voltage of the signal reference (see 9.20). These additional components must be used with great care.

Transients with microsecond rise time and ring frequencies in the kilohertz range, such as the Ring Wave defined by IEEE Std C62.41-1991 [8], are not attenuated rapidly by typical power transformers or building wiring [B31]. Switching of reactive loads, such as transformers and capacitors, create transients in the kilohertz range. Figs 4-2(a) and 4-2(b) illustrate waveforms that are not unusual.

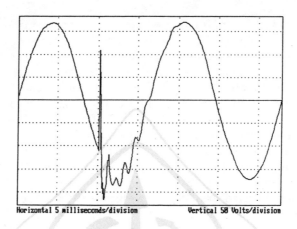

Source: [B10].

**Fig 4-2 (a)
Phase-Neutral Transient Resulting From Addition of Capacitive Load**

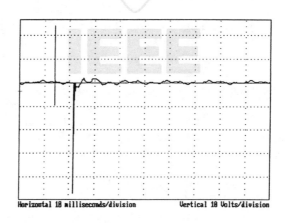

Source: [B10].

**Fig 4-2 (b)
Neutral-Ground Transient Resulting From Addition of Inductive Load**

FUNDAMENTALS

Electromechanical switching devices interact with the distributed inductance and capacitance in ac distribution and loads to create electrical fast transients (EFTs), as shown in Fig 4-3. EFTs are associated with a broad band of frequencies.

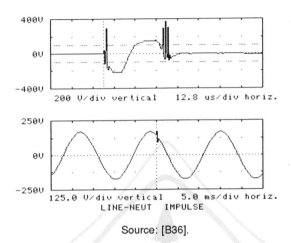

Source: [B36].

**Fig 4-3
Phase-Neutral Transient Resulting From Arcing and Bouncing Contactor**

4.2.3 Building AC Distribution System Impedance. The impedance of local electrical distribution systems is mostly resistive and inductive at power-frequencies of interest (60 Hz to 3 kHz) and mostly inductive and capacitive at higher frequencies (above 1 MHz); see Table 4-1. Therefore, local ac distribution wiring can be used to significant advantage in attenuating unwanted high-frequency noise voltages and short rise-time surges.

Actual impedances of ac feeders and branch circuits vary considerably, due both to their configurations and loads. For purposes of analysis and modeling, equivalent circuits of ac branch circuits have been identified [B14], [B47]. Fig 4-4 depicts the resulting ac branch circuit impedance for such a model reported in [B14]. The general behavior of impedance with frequency, shown in Fig 4-4, is typical for most ac feeder and branch circuits; but actual impedances can vary considerably and resonances above 1 MHz can alter the impedance behavior. It should also be noted that the commonly, but incorrectly, assumed characteristic impedance of 50 Ω for ac distribution circuits can contribute to significant errors, if used to calculate surge energy levels (see 4.5.6).

4.2.4 Load Impedance. Sensitive electronic equipment typically contains motors, transformers, and rectifiers. Outputs of these transformers and rectifiers are typically regulated to provide constant voltage to sensitive load circuits. Insight can be gained as to the nature and operation of these sensitive

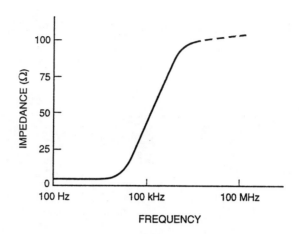

Fig 4-4
Typical AC Distribution Branch Circuit Impedance Versus Frequency
(no load connected)

Table 4-1
Example Cable Impedances at High Frequencies
(copper cable suspended in free air)

(a) #4 AWG Building Wire (25 mm^2)

LENGTH	L (>1 MHz)	@ 1 MHz		@ 10 MHz		@ 100 MHz	
		Rf	$\omega L = Z$	Rf	$\omega L = Z$	Rf	$\omega L = Z$
10 ft	4 µH	0.05 Ω	26 Ω	0.15 Ω	260 Ω	0.5 Ω	2.6 kΩ
20 ft	9 µH	0.1 Ω	57 Ω	0.3 Ω	570 Ω	1.0 Ω	5.7 kΩ
40 ft	20 µH	0.2 Ω	125 Ω	0.6 Ω	1.25 kΩ	2.0 Ω	12.5 kΩ
60 ft	31 µH	0.3 Ω	197 Ω	0.9 Ω	1.97 kΩ	3.0 Ω	19.7 kΩ
100 ft	55 µH	0.5 Ω	350 Ω	1.5 Ω	3.5 kΩ	5.0 Ω	35.0 kΩ

(b) #4/0 AWG Building Wire (107 mm^2)

LENGTH	L (>1 MHz)	@ 1 MHz		@ 10 MHz		@ 100 MHz	
		Rf	$\omega L = Z$	Rf	$\omega L = Z$	Rf	$\omega L = Z$
10 ft	3.6 µH	0.022 Ω	23 Ω	0.07 Ω	230 Ω	0.22 Ω	2.30 kΩ
20 ft	8 µH	0.044 Ω	51 Ω	0.14 Ω	510 Ω	0.44 Ω	5.10 kΩ
40 ft	18 µH	0.088 Ω	113 Ω	0.28 Ω	1.13 kΩ	0.88 Ω	11.3 kΩ
60 ft	28 µH	0.132 Ω	176 Ω	0.42 Ω	1.76 kΩ	1.32 Ω	17.6 kΩ
100 ft	50 µH	0.220 Ω	314 Ω	0.70 Ω	3.14 kΩ	2.20 Ω	31.4 kΩ

FUNDAMENTALS

loads by analyzing their basic components. The basic components of (passive) load impedance each have a distinct variation with frequency. Resistance, R, ideally does not change with frequency. Therefore, its curve is simply a straight horizontal line, with a magnitude of R above the frequency axis (see Fig 4-5(a)).

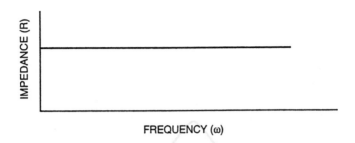

Fig 4-5(a)
Passive Load Resistance Versus Frequency

Inductive reactance, X_L, linearly increases with increasing frequency (of the form $y = mx + b$). Inductive reactance is plotted in Fig 4-5(b), versus frequency, with a slope equal to the inductance, L, of the inductor and intercepting at the origin ($X_L = L\omega + 0$).

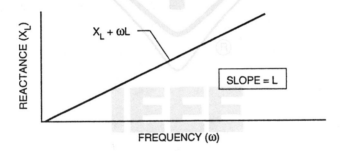

Fig 4-5(b)
Passive Load Inductive Reactance Versus Frequency

Capacitive reactance, X_c, is a hyperbolic function of frequency (of the form $yk = k$), where the frequency, ω, is the independent variable and $1/C$ is the constant. Capacitive reactance versus frequency ($X_c = -1/(\omega C)$) is plotted in Fig 4-5(c). From the above figures it can be seen that, as frequency increases, inductive reactance becomes the dominant factor.

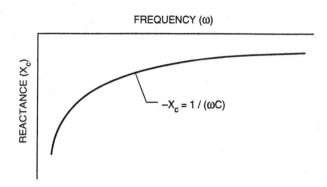

**Fig 4-5(c)
Passive Load Capacitive Reactance Versus Frequency**

4.2.5 Resonance Considerations. AC circuits characteristically have (1) capacitive and inductive elements, and (2) the means to transfer oscillatory energy between these elements. At frequencies where the inductive and capacitive impedances are equal, resonance occurs and the resulting effective impedance can be very high (parallel resonance) or very low (series resonance).

If an ac current source exists at or near the circuit resonant frequency, the circuit voltage at that resonant frequency can rise significantly, especially with little or no resistive load to provide damping. If the circuit is a grounding conductor, it will be effectively open-circuited (not grounded) at the resonant frequencies. It is also possible for current to rise. The voltage and/or current rise is dependent upon the type of resonant circuit and where in the circuit these parameters are being measured.

It is important to analyze the power system frequency response, with the objective of avoiding resonance problems. Resonant frequencies of typical (sensitive electronic equipment) power systems tend to occur in the performance frequency range as defined by 4.2.1.

4.2.5.1 Series Resonance. Series resonance results from the series combination of line/transformer inductances, resistance, and capacitor banks on the ac power distribution system. Fig 4-6(a) shows all three reactance elements superimposed on the same impedance *versus* frequency graph. Series resonance occurs at the frequency, ω_0, where $X_L = -X_c$. The minimum circuit impedance also occurs at the resonant frequency, ω_0, and is equal to the resistance, R, of the circuit. Series resonance acts as a low-impedance path for harmonic currents at the tuned frequency of the circuit.

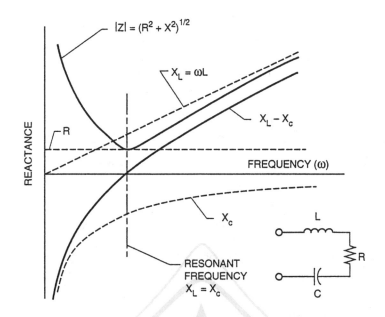

Fig 4-6(a)
Series R-L-C Circuit Impedance Versus Frequency

4.2.5.2 Parallel Resonance. Parallel resonance results from "tank" (LC) circuits in the ac distribution system being excited at frequencies where their inductive and capacitive impedances are equal. Such parallel resonant paths represent very high impedances for currents at their resonant frequency, and can create voltage-breakdown conditions on conductors and components within, or connected to, the circuit. Harmonic currents at the resonant frequency create conditions of high-harmonic voltage across the circuit's terminals, which are also connected to the ac source and its load(s). This frequency-dependent harmonic voltage adds algebraically to the fundamental frequency voltage and to other harmonic voltage waveforms on the circuit, to produce a harmonic distortion of the voltage waveform.

Parallel resonant circuits behave inversely to the series resonant circuit. They exhibit very high impedance at resonance, whereas the series resonant circuit exhibits a very high admittance (low impedance). A diagram of parallel resonance, Fig 4-6(b), appears similar to the series resonance diagram, Fig 4-6(a), when voltages are replaced by currents, currents replaced by voltages, and associated parameters are interchanged with their "inverse equivalents" [B16]. The total set of terms utilized in Fig 4-6(b) and their equivalent series resonance terms are as follows:

(1) Current (I) <=> Voltage (V)
(2) Admittance (Y) <=> Impedance (Z)
(3) Conductance (G) <=> Resistance (R)
(4) Susceptance (B) <=> Reactance (X)
(5) Capacitance (C) <=> Inductance (L)

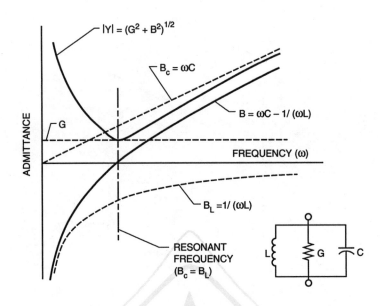

**Fig 4-6(b)
Series R-L-C Circuit Impedance Versus Frequency**

Considerable current can oscillate between the inductive and capacitive storage elements of the circuit when nonlinear loads, with a characteristic harmonic near the parallel resonant frequency, exist in the circuit. Voltage distortion results from these high oscillating current levels. Under certain conditions the oscillating currents can also emit electromagnetic energy, which can interfere with adjacent signal circuits.

Series-resonant circuit currents oscillate though the load and the source, while parallel-resonant circuits confine such current to the parallel circuit's own loop. Therefore, series resonant circuits involve the supply and load with current at the resonant frequency, and parallel resonant circuits impress voltages (at the resonant frequency) on their source and load. This is the underlying mechanism for the production of harmonic voltage waveform distortion.

4.2.5.3 Conductor Self-Resonance Effects. Resonance occurs in conductors, primarily due to their lack of lumped capacitive and lumped inductive elements. Therefore, the conductors of ac electrical distribution systems, which exhibit only distributed capacitance and inductance, oscillate when excited by certain voltage waveforms. They, in essence, act as inadvertent antenna. This type of problem is not often observed in conductors that make up a crude transmission line, such as an ac system feeder or branch circuit; but is a major concern on externally installed grounding/bonding conductors and data cables that form open-loop areas.

FUNDAMENTALS

Conductor self-resonance occurs when a conductor's length equals an odd multiple of 1/4-wavelength of an impressed voltage waveform. The conductor ceases to conduct current at its particular resonant frequency. Current conductance at other frequencies, sufficiently different than resonant frequencies, are not affected ANSI/NFPA 75-1992 [3]. Fig 4-7 depicts this relationship.

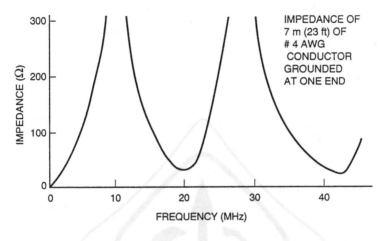

Source: Based on FIPS PUB 94 [11].

**Fig 4-7
Resonance Characteristics of Conductors**

The relationship between the resonant conductor length and frequency is as follows:

$$L_{resonance} = (cn)/(4f_{resonance}) \qquad \text{(Eq 4-1)}$$

where
 $L_{resonance}$ = resonant conductor length (m)
 n = any odd integer (1, 3, 5, ...)
 c = speed of light in free space ($3 \cdot 10^8$ m/s)
 $f_{resonance}$ = frequency of excitation in conductor (Hz)

In practice, designers must be concerned with the lowest frequency at which a given grounding/bonding conductor length will resonate ($n = 1$). Therefore, grounding/bonding conductors should always be chosen so as to not have physical lengths approaching resonant conditions of near quarter-wavelength and odd-multiples thereof for any electrical noise frequencies which might be imposed on the conductor. One hedge against this is to employ multiple grounding/bonding conductors between the same points, each of a different length than the other. Thus, while one path may be undergoing resonance conditions, one or more of the others will not. Usually a difference of 20% in conductor lengths will suffice (FIPS PUB 94 [11]).

4.3 Utility Level Distribution Voltage Disturbances. Electrical utilities in the United States adhere to ANSI C84.1-1989 [1] for the delivery of electrical power. This ANSI document provides guidelines for steady-state tolerances, as shown in Table 4-2.

Continuity of power service at a given site can generally be obtained from utilities. Most utilities have available standard power reliability indices such as ASAI (average service availability index) [B11].

These indices do not include the very short duration interruptions (momentary interruptions). Momentary interruptions generally are considered to be less than a few minutes. They are the result of a variety of normal and abnormal operations in the utility distribution system. Due to the definition of a power interruption generally used by utilities versus the capabilities of sensitive electronic equipment, distribution circuits that the electric utility might consider to be reliable may be totally inadequate to the user of sensitive electronic load equipment. It is advisable that users of sensitive electronic equipment work with their local utility to determine operating characteristics of the particular distribution circuits in question, considering both the frequency of momentary interruptions and pertinent reliability indices.

Voltage waveform disturbances at the utility feeder level have been monitored ([B1], [B11]), and compared and contrasted [B45]. The general conclusion is that line voltage sags are most frequent, and thus most likely to contribute to sensitive electronic load disruptions. They are followed by surges, interruptions, and swells, in lesser probabilities. The actual percentage of each type of voltage disturbance varies with time, location, and the threshold settings of the monitoring instrument. These variances are highly influenced by the particular threshold settings utilized on the monitoring equipment. Appendix A provides an expanded discussion of utility feeder level power quality.

User equipment residing near locations where lightning enters the utility distribution system, will experience high-energy surge conditions via its building ac distribution system. But user equipment located at sites far away from the strike location often experience momentary sag conditions. The momentary sags result from insulator flashover or current shunting actions of the utility's lightning protection equipment which load down the utility's ac distribution system during operation.

4.4 Load and Power Source Interactions. Interactions of interest between sensitive electronic equipment, their power sources, and their environment, primarily result in transient disturbances or steady-state distortions to the system voltage waveform. Table 4-3 summarizes these sources of voltage waveform disturbances and waveform distortions, and their characteristics. When ameliorating these variances, it is often helpful to know their related current waveforms. Certain source/load interactions, e.g., switching, result in short rise-time voltage transients (surges). The reader is referred to 4.5 for a discussion of surges.

Standard Nominal System Voltages and Voltage Ranges

(Preferred system voltages in bold-face type)

VOLTAGE CLASS	NOMINAL SYSTEM VOLTAGE (Note a)			Nominal Utilization Voltage (Note f) Two-wire Three-wire Four-wire	VOLTAGE RANGE A (Note b)			VOLTAGE RANGE B (Note b)		
					Maximum Utilization and Service Voltage (Note c)	Minimum Service Voltage	Minimum Utilization Voltage	Maximum Utilization and Service Voltage	Minimum Service Voltage	Minimum Utilization Voltage
	Two-wire	Three-wire	Four-wire							
Low Voltage (Note 1)	120	**120/240**		115 115/230	126 126/252	**Single-Phase Systems** 114 114/228	110 110/220	127 127/254	110 110/220	106 106/212
						Three-Phase Systems				
			208Y/120 (Note d) **240/120**	**200** 230/115	**218Y/126** 252/126	**197Y/114** 228/114	**191Y/110** 220/110	**220Y/127** 254/127	**191Y/110** (Note 2) 220/110	**184Y/106** (Note 2) 212/106
		240		230	252	228	220	254	220	212
			480Y/277	460	**504Y/291**	**456Y/263**	**440Y/254**	**508Y/293**	**440Y/254**	**424Y/245**
		480		460	**504**	**456**	**440**	**508**	**440**	**424**
		600 (Note e)		575	**630** (Note e)	**570**	**550**	**635** (Note e)	**550**	**530**
Medium Voltage		**2 400**			2 520	2 340	2 160	2 540	2 280	2 080
			4 160/2 400		**4 370/2 520**	**4 050/2 340**	**3 740/2 160**	**4 400/2 540**	**3 950/2 280**	**3 600/2 080**
		4 160			**4 370**	**4 050**	**3 740**	**4 400**	**3 950**	**3 600**
		4 800			5 040	4 680	4 320	5 080	4 560	4 160
		6 900			7 240	6 730	6 210	7 260	6 560	5 940
			8 320/4 800		8 730/5 040	8 110/4 680		8 800/5 080	7 900/4 560	
			12 000/6 930		12 600/7 270	11 700/6 760		12 700/7 330	11 400/6 580	
			12 470Y/7 200		**13 090Y/7 560**	**12 160Y/7 020**	(Note f)	**13 200Y/7 620**	**11 850Y/6 840**	(Note f)
			13 200Y/7 620		**13 860Y/8 000**	**12 870Y/7 430**		**13 970Y/8 070**	**12 504Y/7 240**	
			13 800Y/7 970		14 490Y/8 370	13 460Y/7 770		14 520Y/8 380	13 110Y/7 570	
		13 800			**14 490**	**13 460**	**12 420**	**14 520**	**13 110**	**11 880**
			20 780Y/12 000		21 820Y/12 600	20 260Y/11 700		22 000Y/12 700	19 740Y/11 400	
			22 860Y/13 200		24 000Y/13 860	22 290Y/12 870		24 200Y/13 970	21 720Y/12 540	
		23 000			24 150	22 430	(Note f)	24 340	21 850	(Note f)
			24 940Y/14 400		**26 190Y/15 120**	**24 320Y/14 040**		**26 400Y/15 240**	**23 690Y/13 680**	
			34 500Y/19 920		**36 230Y/20 920**	**33 640Y/19 420**		**36 510Y/21 080**	**32 780Y/18 930**	
		34 500			36 230	33 640		36 510	32 780	
		46 000			Maximum Voltage (Note g) { 48 300					
		69 000			72 500					
High Voltage		**115 000**			121 000					
		138 000			145 000					
		161 000			169 000					
		230 000			242 000					
		(Note h)								
Extra-High Voltage		**345 000**			362 000					
		500 000			550 000					
		765 000			800 000					
Ultra-High Voltage		1 100 000			1 200 000					

NOTES: (1) Minimum utilization voltages for 120–600 volt circuits not supplying lighting loads are as follows:

Nominal System Voltage	Range A	Range B
120	108	104
120/240	108/216	104/208
(Note 2) 208Y/120	187Y/108	180Y/104
240	216	208
480Y/277	432Y/249	416Y/240
480	432	416
600	540	520

(2) Many 220 volt motors were applied on existing 208 volt systems on the assumption that the utilization voltage would not be less than 187 volts. Caution should be exercised in applying the Range B minimum voltages of Table 1 and Note (1) to existing 208 volt systems supplying such motors

Source: ANSI C84.1-1989 [1].

Table 4-3
Matching Sensitive Load and Power Source Requirements With Expected Environments

Voltage Parameter Affecting Loads	Typical Range of Power Sources	Typical Immunity of Electronic Loads		
		Normal	Critical	Units affected and comments
Over and Undervoltage	+6%, -13.3%	+10%, -15%	± 5%	Power supplies, capacitors, motors. Component overheating and data upset.
Swells/Sags	+10%, -15%	+20%, -30%	± 5%	Same as above.
Transients, Impulsive & Oscillatory, Power Lines	Varies: 100-6000 V	Varies: 500-1500 V	Varies: 200-500 V	Dielectric breakdown, voltage overstress. Component failure and data upset.
Transients, Impulsive & Oscillatory, Signal Lines	Varies: 100-6000 V	Varies: 50-300 V	Varies: 15-50 V	Same as above.
ESD	<45 kV 1000-1500 V	Varies widely 200-500 V	Varies widely 15-50 V	Signal circuits. Dielectric breakdown, voltage overstress, Component failure, data upset. Rapid changes in signal reference voltage.
RFI/EMI (Conducted) (normal and common mode)	10 V up to 200 kHz less at higher freq.	Varies widely 3 V typical	Varies widely 0.3 V typical	Signal circuits. Data upset, rapid changes in signal reference voltage.
RFI/EMI (Radiated)	<50 kV/m, <200 kHz <1.5 kV/m, >200 kHz	Varies widely w/ shielding	Varies widely w/ shielding	Same as above.
Voltage Distortion (from sine wave)	5-50% THD	5-10%	3-5%	Voltage regulators, signal circuits, capacitor filters, capacitor banks. Overheating, under-charging.
Phase Imbalance	2-10%	5% max	3% max	Polyphase rectifiers, motors. Overheating.

Current Parameter Affecting Sources	Typical Range of Load Current	Typical Susceptibility of Power Sources		
		Normal	Critical	Units affected and comments
Power Factor	0.85-0.6 lagging	0.8 lagging	<0.6 lagging or <0.9 leading	Power source derating or greater capacity source with reduced overall efficiency.
Crest Factor	1.4-2.5	1.0-2.5	>2.5	1.414 normal; impact function of impedances at 3rd and higher harmonics (3-6% Z). Voltage shape distortion.
Current Distortion	0-10% total rms	5-10% total 0-5% largest	5% max total 3% largest	Regulators, power circuits. Overheating.
DC Current	Negligible to 5% or more	<1%	As low as 0.5%	Half-wave rectifier loads can saturate some power sources, trip circuit breakers.
Ground Current	0-10 A rms + noise and surge currents	>0.5 A	<0.1 A	Can trip GFI devices, violate code, cause rapid signal reference voltage changes.

Frequency Parameter Affecting Loads	Typical Range of Power Sources	Typical Immunity of Electronic Loads		
		Normal	Critical	Units affected and comments
Line Frequency	± 1%	± 1%	± 0.5%	Zero-crossing counters.
Rate of Freq. Change	1.5 Hz/s	1.5 Hz/s	0.3 Hz/s	Phase sychronization circuits.

Source: Based on FIPS PUB 94 [11].

4.4.1 Transient Voltage Disturbance Sources/Characteristics. Here we consider voltage waveform disturbances to be that set of voltage variances on the power circuits of interest, which are (1) nonsinusoidal at the nominal frequency of the power source, and (2) the result of power source and load characteristics, as interacted by the ac building distribution system. These disturbances in system voltage waveform tend to decay rapidly with time.

Load-related changes and switching events cause almost all voltage disturbances that occur between sensitive equipment and their power sources. Several common load-derived sources of voltage waveform disturbances and their relative characteristics are presented below.

4.4.1.1 Step Loads. Step load changes are one of the most common sources of voltage disturbance. The basic cause of the voltage disturbance is simply the voltage drop caused by the load current and the power system impedance. Simply stated, when load current changes, the voltage drop changes. Voltage regulators tend to correct voltage drops within the power distribution system, but only after a time delay that is an inherent characteristic of the feedback regulator utilized.

4.4.1.2 Inrush Currents (Motors, LC Line Filters, and Power Supplies). Inrush currents, associated with the initial energizing of motors, low-pass type LC line filters, and power supplies, are typically found in sensitive electronic equipment.

AC motor starting (inrush) currents are typically equal to the locked-rotor currents, which are typically 5 to 7 times their rated full-load current. These inrush currents can require typically 0.3 to 3.0 s to decay to steady-state values, depending on acceleration time. DC motor starting currents appear as rectifier loads on the ac power distribution system.

The initial energizing of transformers creates (magnetizing) current transients. Inrush currents 10 to 20 times their normal full-load current can exist, decaying in several cycles under worst-case conditions. Actual inrush currents will depend on the phase angle of the initial voltage waveforms and the state of residual magnetic (core) flux from prior transformer energizing. When rectifier/capacitor power supplies are energized, the initial capacitor charging can cause similar levels of current inrush.

AC/DC power supplies often have a large filter capacitor bank charged as much as possible on the first half-cycle of the applied current. In these cases the charging current is often limited only by circuit impedances.

4.4.1.3 Fault Currents. Fault currents represent an extreme case of transient current flow and thus ac line voltage disturbance. Depending on the power system impedance, several orders of magnitude of normal full load current may be available. Severe voltage reductions to adjacent equipment usually result until the fault is cleared. Some fault conditions do not result in high currents and may not cause overcurrent protective devices to operate (e.g., arcing ground faults, under certain conditions). These faults often create significant high-frequency transient voltage of large amplitude. Solidly grounded power sources tend to minimize this type of fault.

4.4.1.4 Voltage Regulator Interactions. Electronic loads are typically equipped with internal voltage regulators. If poorly applied, these internal voltage regulators may negatively interact in the ac distribution. The result can range from a tendency to amplify ac line voltage disturbances to uncontrolled oscillation of the input ac voltage to the sensitive load (FIPS PUB 94 [11]).

4.4.2 Potential Impacts of Transient Voltage Disturbances. Disturbances of the ac voltage waveform and their attendant current harmonics have been shown to significantly impact both the ac distribution system and the sensitive electronic loads (FIPS PUB 94 [11]). The most significant of these are discussed below.

4.4.2.1 Complete Loss of AC Power to Electronic Loads. Excessive motor and transformer inrush currents can exceed the time-current trip curves of upstream overcurrent protection devices, causing electronic loads to open circuit.

4.4.2.2 Short-term Voltage Variances. Temporary reductions in the ac distribution voltage can be caused by significant step changes in load current. This is particularly true for transformer and motor inrush currents, and large sensitive load systems that dynamically switch on/off their subsystems (FIPS PUB 94 [11]). Time duration of these low ac voltages can exceed the hold-up time of sensitive load dc power supplies, thus causing the equivalent of an ac line voltage sag or interruption.

4.4.2.3 Transient Phase Shift Due to Reactive Load Changes. This effect is primarily the result of dynamic switching (on/off) of inductive and capacitive load elements (e.g., ac motors and shunt capacitors). These large dynamic changes in load current, fed by reactive ac circuits, result in voltage time-shifts on the ac circuits.

4.4.2.4 Data Upset. Many of the aforementioned disturbances may occur with no other effect on the connected sensitive electronic load equipment except to inadvertently activate internal equipment, power quality checking circuits, and to trigger them into alarm or error status.

4.4.2.5 Frequency Variations and Slew-Rate. When an on-site generating system, such as an engine-alternator is used as the ac power source for sensitive electronic load equipment, variations in loading can cause variations in rotational speed, thus frequency shifts occur. Additionally, a rate-of-change of frequency between zero-crossing points, i.e., slew-rate, may occur. High slew-rate conditions often create problems within sensitive load equipment.

4.4.3 Steady-state Voltage Distortion Sources/Characteristics

4.4.3.1 Nonlinear Loads. When the instantaneous load current is discontinuous or is not proportional to the instantaneous ac voltage, it is termed nonlinear. The effect is equivalent to the presence of harmonic (higher frequency) components of current superimposed upon the nominal (e.g., 60 Hz) sinusoidal current (ANSI/NFPA 75-1992 [3]). All components added together equal the actual current waveform. These components of current are not in phase with the distribution voltage waveform (at each harmonic frequency). These harmonic currents also interact with the power source impedance and typically create voltage distortion, excite power system resonances, and stress power system components on the ac distribution system. These disturbances and proposed limits are discussed in detail in IEEE Std 519-1992 [B27].

Many sensitive electronic loads exhibit nonlinear characteristics. AC/DC power supplies using simple diode rectifiers and dc filter capacitors are common examples of this type of load. They are often used by manufacturers of sensitive electronic equipment. More sophisticated ac/dc converters with improved power factor and greatly reduced harmonic currents are becoming available, primarily as a result of the harmonic current limits required by [B21], [B22], and [B23].

Exact analysis of power supplies is complex, but it can be said that a load current flows nonlinearly during the ac cycle (ANSI/NFPA 75-1992 [3], [B3]). The duration of current flow (each half cycle, on each phase) can be described in terms of the conduction angle. Theoretically, the conduction angle varies between 0 and 180 degrees (1/2 cycle), and varies with load current and ac line voltage. Typical conduction angle for switch-mode power supplies is 30–60 degrees, and typical current crest-factors range from 2 to 3 (versus 1.4 for a linear load fed by sinusoidal ac power).

Table 4-4 shows an example of the harmonic current content of balanced line-to-line and line-to-neutral diode-capacitor power supply in a three-phase power system. In three-phase circuits, the triplen harmonic neutral currents (third, sixth, ninth, etc.) add instead of cancel, since they are multiples of three times the fundamental power frequency and are spaced apart by 120 electrical degrees. (Based on the fundamental frequency, triplen harmonic currents of each phase are in phase with each other, and so add in the neutral circuit.) Under worst-case conditions, the neutral current can be 1.73 times the phase current [B18].

4.4.3.2 Reactive Loads. Sensitive electronic loads can also impact the ac distribution system by causing non-unity power factors. Power supplies, ac motors, low-pass LC filters, and other components within sensitive loads often cause both non-unity displacement power factor and distortion power factor.

Table 4-4
Example Input Harmonic Current Distortion in Balanced Three-Phase Circuits Due to Rectifier-Capacitor Power Supply

Harmonic Number	Line-to-Line Harmonic Current*	Line-to-Neutral Harmonic Current*
1	0.82	0.65
3	—	0.52
5	0.49	0.42
7	0.29	0.29
9	—	0.13
11	0.074	0.12
13	0.033	0.098
Total phase current	1.00	1.00
Neutral current	0.0	1.61

* Normalized to phase current.

4.4.4 Potential Impacts of Steady-state Current Distortions

4.4.4.1 Transformer Eddy Current Heating Due to Harmonic Currents. Transformers serving nonlinear (typically electronic) loads exhibit increased winding (eddy current) losses due to harmonic currents generated by those loads.

4.4.4.1.1 Derating Conventional Transformers. Conventional power transformers typically require derating when serving nonlinear loads. ANSI C57.12.00-1987 [6] (for liquid-immersed) and IEEE Std C57.12.01-1989 [7] (for dry-type) specify the following restrictions to obtain full rated transformer performance:
  (1) Approximately sinusoidal, balanced voltage.
  (2) Load current that does not exceed 0.5% total harmonic distortion (THD).

These limitations are primarily due to harmonic current induced eddy currents in the windings that increase losses and can cause overheating. Voltage harmonics can also cause additional losses in the core, but in most practical cases the harmonic current-related winding losses are the limiting factor for transformer capacity. Skin effect also plays a role at higher frequencies and large diameter conductors, but is not considered in most practical 60 Hz power system applications.

The recommended practice for derating conventional transformers in applications where nonsinusoidal load currents are present is provided in [B25]. The standard applies the results of studies that found winding eddy-current loss, Pec, to be approximately proportional to the square of the rms load current at that harmonic, I_h, and the square of harmonic number, h [B9].

If one knows the eddy current loss under rated conditions for a transformer, (Pec-r), the eddy-current loss due to any defined nonsinusoidal load current (Pec) can be expressed as follows [B25]:

$$\text{Pec} = \text{Pec-r} \sum_{h=1}^{h=h_{max}} I_h^2 \cdot h^2 \qquad \text{(Eq 4-2)}$$

where
I_h = rms current at harmonic h
h = harmonic order

This relationship has been found to be more accurate for lower harmonics (3rd, 5th, 7th), and an overestimation of losses for higher harmonics (9th, 11th, . . .), particularly for large diameter windings and large kilovoltampere transformers [B12], [B19].

4.4.4.1.2 K-Factor Rated Transformers. Underwriters Laboratories (UL) and transformer manufacturers established a rating method called K-factor, for dry-type power transformers to indicate their suitability for nonsinusoidal load currents. This K-factor relates transformer capability to serve varying degrees of nonlinear load without exceeding the rated temperature rise limits. This K-factor is based on predicted losses as specified in the simplified method of [B25]. The limiting factor related to overheating is again assumed to be eddy current losses in the windings. The K-factor is defined on a per unit basis as follows [B2], [B48]:

$$K = \sum_{h=1}^{h=h_{max}} I_h^2 \cdot h^2 \qquad \text{(Eq 4-3)}$$

where
I_h = rms current at harmonic h, in per unit of rated rms load current

The same factor can be seen in (Eq 4-2). For rating purposes UL has specified that the rms current of any single harmonic greater than the tenth harmonic be considered as no greater than $1/h$ of the fundamental rms current. This limitation is an attempt to compensate for otherwise overly conservative results at higher harmonic frequencies.

Standard K-factor ratings are 4, 9, 13, 20, 30, 40, and 50. The current in Eq 4-3 is expressed on a per-unit basis such that the sum of the individual currents times the harmonic number squared is 1. Thus for a linear load current, the K-factor is always one. For any given nonlinear load, if the harmonic current components are known, the K-factor can be calculated and compared to the transformer's nameplate K-factor. As long as the load K-factor is equal to, or less than, the transformer K-factor, the transformer does not need to be derated. An example nonlinear load's K-factor is shown in Table 4-5. UL lists the K-factor nameplate rating for dry-type transformers under [B2] and [B48]. Testing with a nonlinear load of appropriate K-factor is the preferred method. However, due to practical limitations, the most common method employs an overload of fundamental load current to simulate harmonic loading. This test method is described in the UL Standards cited and requires an adjustment to compensate for harmonic losses.

Table 4-5
Example Calculation of a Nonlinear Load's K-Factor

Harmonic Number h	Nonlinear Load Current I_h	I_h^2	i_h *	i_h^2	$i_h^2 \cdot h^2$
1	100%	1.000	0.909	0.826	0.826
3	33%	0.111	0.303	0.092	0.826
5	20%	0.040	0.182	0.033	0.826
7	14%	0.020	0.130	0.017	0.826
9	11%	0.012	0.101	0.010	0.826
11	9%	0.008	0.083	0.007	0.826
13	8%	0.006	0.070	0.005	0.826
15	7%	0.004	0.061	0.004	0.826
17	6%	0.003	0.053	0.003	0.826
19	5%	0.003	0.048	0.002	0.826
21	5%	0.002	0.043	0.002	0.826
Total		1.211		1.000	9.083
				K-factor =	9.083

*$i_h = I_h / (\Sigma I_h^2)^{1/2}$

4.4.4.2 Triplen Harmonic Load Generated Overcurrents in Wiring. In three-phase, four-wire circuits, currents associated with the triplen (odd multiples of three) harmonics which algebraically add in the neutral conductor, have in extreme cases resulted in fires. This is a particular problem in that neutral conductors are not subject to the normal overcurrent protection in ac distribution systems (ANSI/NFPA 75-1992 [3], [B24]).

4.4.4.3 Resonance Due to Harmonic Load Currents. The presence of capacitors, such as those used for power factor correction, can result in parallel resonance which causes excessive currents with subsequent damage to such capacitors as well as excessive ac line voltage distortion (FIPS PUB 94 [11]).

4.4.4.4 Heating Due to Nonsinusoidal Voltage Source. Depending on the impedance of the power source, nonlinear loads will cause nonsinusoidal

FUNDAMENTALS

voltage waveforms. Voltage supplied to other equipment (e.g., ac motors and transformers) with these distorted waveforms can result in additional heat dissipation (ANSI/NFPA 75-1992 [3]).

Harmonic currents flowing because of source voltage distortions typically cause significant heating in ac motors. Stator windings, rotor circuits, and stator and rotor laminations tend to absorb additional energy due primarily to eddy currents, hysteresis, and to a lesser degree, skin effect. Leakage fields set up by harmonic currents in stator and rotor end-windings also produce extra losses [B3].

4.4.4.5 Phase Shift (Power Factor) Effects. The total power factor is the combination of the displacement and distortion power factors. Total power factors of sensitive electronic loads rarely approach unity (ANSI/NFPA 75-1992 [3], [B3]), and always should be a concern of the ac distribution designer.

Distortion power factor accounts for the flow of reactive (harmonic) power where the load current and ac line voltage are neither sinusoidal nor of the same frequency. Nonlinear loads act as generators of these (harmonic) currents which are imposed on the power source.

4.4.4.6 Subcycle Voltage Waveform Variances. Nonlinear loads exhibiting large crest-factors tend to cause flat-topping of ac distribution voltage waveforms. These large crest-factors can preclude certain types of sensitive load dc power supplies from obtaining needed current from the building ac power distribution system. This excessive flat-topping can cause the equivalent of a power line sag (FIPS PUB 94 [11]).

Thyristor-controlled loads (e.g., power supplies, motor controls) can cause ac distribution voltage disturbances such as notching and multiple zero-crossings. These disturbances in turn can upset sensitive electronic loads.

4.4.5 Corrective Means. Several means of control for ac distribution voltage disturbances are possible. They can be grouped into the following five categories:

(1) Sensitive load(s) characteristics;
(2) Interface equipment (e.g., power conditioners) characteristics;
(3) Building ac distribution and grounding circuits characteristics;
(4) Power source (e.g., utility, on-site generator) characteristics;
(5) Power factor means within the building ac distribution system.

Often in practice, only techniques for reducing voltage disturbances are within the control of the ac distribution system designer. Building level ac distribution modifications and interface devices, such as power conditioners, can be appropriate. Chapter 9 identifies detailed design guidelines for the reader.

4.5 Voltage Surges. Voltage surges are subcycle voltage transients. They are a particular concern for sensitive electronic equipment. Very small voltage surges have been documented to cause disruption of data flow and integrity (FIPS PUB 94 [11]). Higher energy voltage surges are often responsible for the destruction of components within the sensitive equipment [B13], [B49].

4.5.1 Sources/Characteristics. There exist a large number of potential sources of electrical surges that can cause harm to sensitive electronic equipment and systems. The majority of these sources can be divided into two major categories: circuit switching and environmental.

4.5.1.1 Switching Surges. Switching surges are associated with rapid changes in current flow rates (di/dt) within a given electrical system. Physically, one can visualize switching surges as being the expansion or reduction of magnetic/electric fields.

Decay rates of these induced voltages are generally slower than their rise rates, and are long relative to power system time-constants. Switching surges can take several forms depending on system configuration and rate of change in operating conditions.

Typical causes include the following:
(1) Energizing or deenergizing premises power source or (reactive) load equipment,
(2) Arcing associated with lose connections or ground faults, and
(3) Power factor capacitor switching.

Fig 4-8 depicts a generalized power network with self-inductances, L_L, mutual inductances, L_M, resistances, R, and capacitances, C. Changes in currents with time for all the closed circuits (loops) described by Fig 4-8 can be generally described by Kirchhoff laws. Assuming L_L, L_M, R, and C are constant, the total current flows can be divided into steady-state and transient components. The transient current components are of interest.

These transient currents produce transient fluxes and charge levels within individual components in the circuit. The following results can be shown in ac circuits with resistance, inductance, and capacitance [B41]:
(1) There is no discontinuity in voltage or current at the time of switching.
(2) A decaying alternating current and voltage develops with time.
(3) The magnitude of the voltage disturbance (switching surge) is determined primarily by the initial voltage and circuit capacitance.

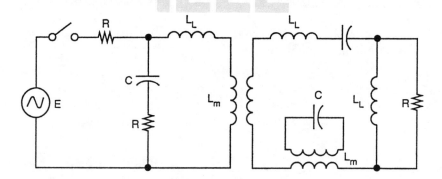

Fig 4-8
Generalized Power Network

FUNDAMENTALS

Applying these concepts to the case of a typical distribution wiring system with a distant short circuit and interrupted by an overcurrent protection device somewhere in the line (depicted in Fig 4-9), we can further state [B41]:

(1) The amplitudes of the transient oscillations are determined by the switching current in the inductance and the switching voltage across the capacitance.
(2) The switching current and voltage change sinusoidally and in general have a phase difference.
(3) Switching surges can attain a theoretical maximum of twice their source voltage.

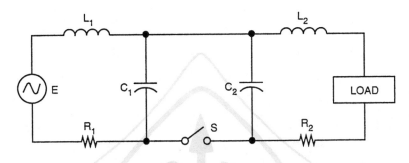

Fig 4-9
Typical AC Building Distribution Wiring System

Fig 4-10 (a) and (b) indicates the general behavior of switching surge voltage and current respectively, with a stable arc drawn between the contacts. Note that before interruption the switching surge (circuit breaker) voltage is zero. Then the surge voltage increases slowly after separation of contacts occurs, followed by damped oscillation.

If the switching arc is unstable (inductive-capacitive circuit) as the contacts open, then the current is often interrupted and reignited several times before the dielectric strength of the increasing contact gap distance overcomes the voltage difference across the gap, thus creating a stable open-circuit condition. Fig 4-3 depicts the surge voltages associated with this multiple interruption-reignition across the switch contacts. It should be noted that the multiple interruption-reignition yields a series of electrical-fast-transient, having a relatively long rise-time ending with an abrupt collapse.

In general, the fast-rising wavefronts of switching surges are slowed by discontinuities in the capacitance and inductance of the building ac distribution system (due primarily to lumped capacitance and inductance) from their point of incidence to the sensitive equipment [B44], and their amplitudes are reduced as a function of the losses in the transmission medium (losses increase with increasing path lengths). That is to say, the closer electrically that the sensitive equipment is to the sources of switching surges, the more severe the potential effect of the transient on that equipment. This attenuating effect

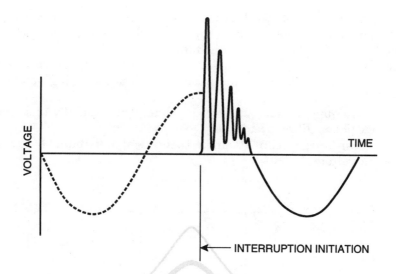

**Fig 4-10(a)
Typical Behavior of (Power-off) Switching Transient (Recovery) Voltage
Without Multiple Interruption-Reignition**

of the building ac distribution system is highly dependent upon the rise-time of the surge. The maximum voltage of a 5 ns rise-time surge is reduced by a factor of two, via 60–70 m of a low-voltage, single-phase distribution branch circuit (in steel conduit) [B34]. Very little voltage attenuation is observed for longer rise-time surges, as reported in [B32]. Transmission line effects, such as reflected voltage-waves (high-frequency energy on long conductors), are dramatically different, thus actual design characteristics and conditions should also be assessed [B8], [B41].

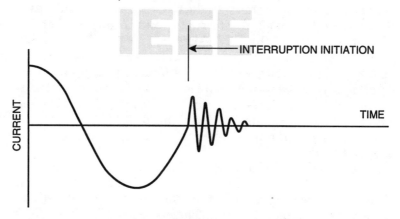

**Fig 4-10(b)
Typical Behavior of (Power-off) Switching Transient (Recovery) Current
Without Multiple Interruption-Reignition**

4.5.1.2 Environmentally Induced Surges. Lightning is the most obvious environmentally generated electrical transient. In addition, non-arcing atmospheric charge redistribution and ground-based electrostatic discharges are significant contributors to data disruption and damage to sensitive equipment [B46].

4.5.1.2.1 Lightning-Induced Surges. Considerable information exists in the literature as to the mechanics of lightning strokes and their formation. Here it suffices to state that the development of large negative charge centers in the lower regions of clouds cause a corresponding positive charge center to be induced on the earth's surface below them. This results in a potential (voltage) between the cloud and earth. Such charge centers continue to develop until the voltage gradient, at the cloud base, exceeds the breakdown strength of air. The result is a low-current discharge, i.e., a pilot streamer. The pilot streamer is immediately followed by a higher current discharge, i.e., a stepped leader, which is followed by one or more (ground-to-cloud) return strokes [B6], [B44].

As many as forty return strokes have been observed [B35]. Their currents range from a few hundred A to more than 500 000 A, as shown in Fig 4-11. The 50th percentile is 20 000 A, and the 90th percentile is approximately 200 000 A. They are relatively fast acting, existing only 50–100 µs. It is also important to note that the return strokes' rise-time is very short, typically 0.1–10 µs.

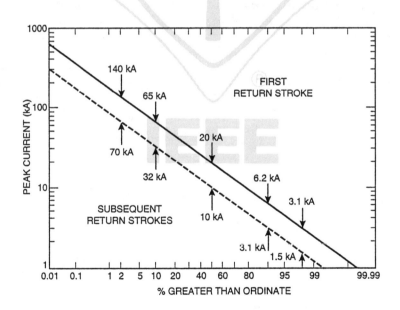

Source: Based on [B8].

**Fig 4-11
Distribution of Lightning Stroke Current**

Their large current levels create an ionized volume in the surrounding earth (ionization region). Within this volume of earth, the lightning energy arcs directly to any highly conductive elements (e.g., buried cables), thus causing voltage rises considerably higher than in areas of earth that are outside the ionization volume. The shape and size of this ionization region is a function of the soil resistivity and the current in the lightning stroke. This region is of particular importance in the suppression of the lightning stroke's impact on electronic equipment and/or conducting cables.

Fig 4-12 shows the arcing distances for bare conductors and for insulated conductors as a function of stroke current and soil resistivity. Note that arcing distances of over 100 m are possible with soils having an electrical resistivity of 1000 Ω·m or greater (observed in several regions of the United States). Outside of this ionization region, the available lightning induced voltage is considerably reduced, and thus the induced voltages into nearby electrical conductors are also lessened.

The parts of the lightning phenomenon most important in the design of lightning protection for sensitive electronic equipment are the latter stage of the stepped leader process and the subsequent high-current return discharges. The most important characteristics of the discharge are its current, voltage, waveshape, polarity, charge, and frequency of occurrence.

Cloud-to-cloud discharges can also induce considerable transient energy into aerial and buried conductors [B6].

4.5.1.2.2 Nonarcing Atmospheric Charge Redistribution. Significant levels of transient energy can be induced into both buried and overhead conductors from the rapid redistribution of atmospheric (cloud) charge centers. This phenomenon commonly occurs after lightning strokes, and is the result of the highly mobile charge centers attempting to find an equilibrium with the relatively fixed earth charges. The rapid movement of charge causes electromagnetic fields similar to those of a cloud-to-cloud stroke. The resulting voltage and current surges in overhead and buried conductors are modeled similarly to cloud-ground lightning strokes, except with an expanded time base [B46].

4.5.1.2.3 Electrostatic Discharge. Electrostatic discharges (ESD) typically have a high reference potential, but low amounts of energy. Several charge generation processes exist, including triboelectrification, induction charging, and corona charging [B16]. Static charge buildup typically results from a "rubbing action" between two materials (solid or liquid) of different surface-energy characteristics, in the absence of a conductive path between them. This buildup of charge is quickly released when a conductive path (discharge arc) is established [B5]. ESD surges can be very harmful to semiconductor devices in sensitive electronic equipment. Discharge voltages are often in the range of 5–40 kV [B13]. Energy levels tend to be of the order of units of mJ to tens of mJ.

One can characterize these surges as having very short rise-times (high dv/dt) and relatively slow decay rates (as compared to lighting or switching

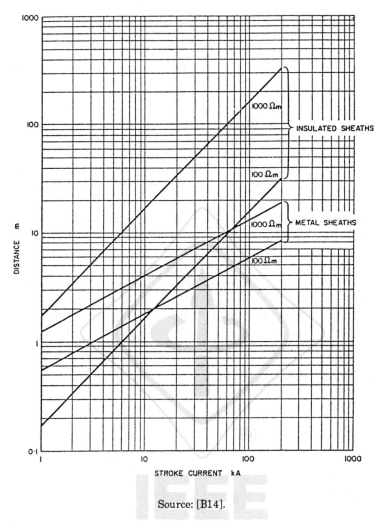

Source: [B14].

**Fig 4-12
Arcing Distances for Bare and Insulated Conductors**

induced surges). Since these surges have little energy, they can be relatively easily negated by the use of (fast responding) voltage clamps and capacitors [B44]. Shielding of sensitive circuits is also an effective means of protection. Due to their very short rise times, ESD surges attenuate considerably within the building ac distribution wiring. Therefore, ESD control is most important for sources that are in close proximity to sensitive electronic circuits.

The most effective ESD control methods include limiting the discharge rate (or path), as well as the rate of charge buildup. ANSI/NFPA 77-1988 [4] should be consulted for detailed design information.

4.5.1.3 Other Sources. In addition to switching and environmentally generated surges, high-altitude nuclear detonation can also generate surges in terrestrial-based sensitive electronics and their power systems. Nuclear detonations generate high-intensity electromagnetic pulses (nuclear EMPs, or NEMPs).

The NEMP of concern to terrestrial electronics is created by an exoatmospheric nuclear explosion. These electromagnetic pulses can affect large regions (due to the height of the blast above the earth) and in general can be considered to have a spherical wave geometry [B37], [B51].

To an observer on the ground, the incoming electromagnetic waves appear to be planar, propagating toward him from a single point. Therefore, unlike lightning, NEMPs affect large geographical areas. This is of particular concern for long, interconnected conductors such as utility distribution systems [B30].

Like lightning, NEMP can produce large electrical transients in smaller installations of sensitive electronic equipment. Peak electric fields can be thousands or tens of thousands of volts per meter, with rise times on the order of 10 ns and decay times an order of magnitude of more longer [B37]. These very high rates of change of field intensity also aid the coupling of NEMPs into terrestrial electrical and electronic systems. Protection schemes typically require a combination of shielding and series-parallel combinations of transient voltage surge suppression (TVSS) devices. The reader is referred to [B37] for a more detailed discussion of NEMPs.

4.5.2 Coupling Mechanisms. Coupling mechanisms for short rise-time surges can be categorized as free-space or far-field.

4.5.2.1 Free-Space Coupling. Free-space coupling, i.e., stray reactive coupling, of surge energy can be divided into two ranges, near-field and far-field. The near-field range involves those conductors which are of sufficiently close proximity that inductive (magnetic) or capacitive coupling can occur. Far-field coupling of surge energy involves the radiation (and interception) of electromagnetic waves as the principal coupling mechanism.

4.5.2.1.1 Inductive (Magnetic) Coupling (Near Field). Sensitive electronic circuits that are physically near, but not in direct contact with, a surge path can experience damage without flashover (discharge) occurring. Due to the high di/dt characteristic of surges, voltages can be electromagnetically induced on nearby conductors. This effect is depicted in Fig 4-13, for the case of surge current on the down-conductor of a lightning interception system. Voltage induced into the adjacent circuit (loop) is a function of the di/dt of the surge current traveling through the down-conductor. Actual voltages induced are a function of the loop geometry, distance from the down-conductor, and the time rate-of-change of the surge current. Fig 4-14 plots normalized induced voltage per unit length (l) developed in a sensitive circuit having various loop geometries.

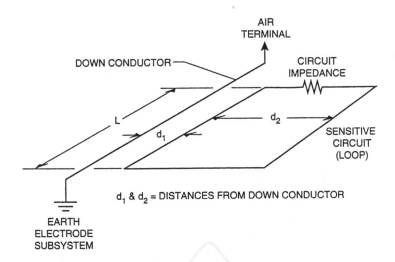

Source: Based on [B37] and [B38].

**Fig 4-13
Inductive Coupling of Surge Current to Adjacent Sensitive Circuits**

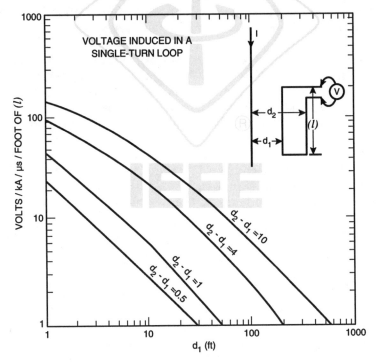

Source: Based on [B37] and [B38].

**Fig 4-14
Normalized Induced Voltage Into Sensitive Circuits**

This general unwanted coupling of surge energy into sensitive circuits is possible whenever any set of similar conductors, with one carrying the initial surge, are in close proximity. The loop area enclosed by the sensitive circuit is the important parameter, i.e., more area means more coupling problems. Coupling can also be reduced by positioning the sensitive circuits at right angles to local surge-source circuits.

The voltages magnetically induced into sensitive circuits are not a function of the sensitive circuit impedance, therefore the magnitudes of induced voltages are the same for low- and high-impedance circuits. This fact can be particularly problematic for low-impedance circuits, and any circuit without surge protection.

4.5.2.1.2 Capacitive (Electrostatic) Coupling (Near Field). Capacitive coupling of surge energy, an electrostatic phenomenon, can occur between sufficiently close circuits. Parameters of interest are spacing, mutually exposed area, and dielectric constant between the source and the sensitive circuit [B40]. Coupling levels are dependent on the amplitude and rate-of-change of the voltage surge, coupling capacitance, stray capacitance in the sensitive circuit, and impedance of the sensitive circuit. Low-frequency capacitively coupled voltages increase with increasing coupling capacitance and with increasing sensitive circuit impedance. At higher frequencies, capacitively coupled voltages increase with increasing coupling capacitance and with decreasing stray capacitance in the sensitive circuit.

4.5.2.2 Far-Field (Electromagnetic) Coupling. For sensitive (i.e., victim) circuits, far-field coupling of electromagnetic energy occurs when the sensitive circuit acts as a receiving antenna for incident electromagnetic energy [B4]. Currents induced in this manner increase with the strength of the electromagnetic field in the vicinity of the sensitive circuit, and with the circuit effectiveness as a receiving antenna. The field strength is an inverse function of the distance from the radiating source. The sensitive circuit effectiveness as an antenna depends on its length (and geometry) relative to the wavelength(s) of the noise signal(s). The sensitive circuit must exhibit the necessary conditions of resonance at the interfering field's frequencies.

4.5.3 Interaction With Buried Cables. Analysis and measurements of transient voltages induced into buried cables [B6],[B46] indicate that surges are a function of the cable parameters, depth of the cable(s), soil resistivity, cable terminations, and the additional degree of shielding provided by buildings, water pipes, power lines, and other nearby conductors.

Cable parameters of importance are the cable length, the "transfer impedance" of the cable shield, and the dielectric strength of the insulating jackets [B39]. Soil resistivity is also important in determining the surges induced by lightning. References [B39] and [B47] indicate that the peak transient voltages and currents are approximately proportional to the square root of the soil resistivity.

Deeply buried cables generally suffer less from the direct effects of lightning strokes, due to greater attenuation of the surge's higher frequencies near the earth's surface. Similarly, guard-wires above buried cables can be effective in reducing the impact of ground currents.

4.5.4 Interaction With Above-Ground Conductors. The use of aerial conductors to intercept lightning strokes and protect cables below them from the direct effects of lightning has been well demonstrated. Several theories have been developed to explain the size of the protected zone. They are reviewed in [B14]. Use of these concepts can reduce both the voltage and current surge levels that above-ground power and signal lines experience, for a given lightning stroke.

Lightning generated surges on utility ac distribution systems, and at the user's site, have been studied extensively and have been reported in the literature. References [B14] and [B50] are examples of these studies. References [B14] and [B28] also provide a history and bibliographies of the problem and a summary of measurements and operating experiences pertinent to remote ac distribution lines. The types of damage observed and the surges measured at distribution terminals are also discussed. Protection strategies for terminal equipment have been well developed and consist of surge current diverters and/or grounded overhead guard-wires.

Elevated conductors (ac distribution, etc.) form geometric loops of various sizes and orientations. As a general statement, the open-circuit voltages induced in these loops are a function of loop size and the time-rate-of-change of the magnetic flux through the loop cross-sectional area [B14]. Therefore, the peak open-circuit voltage is dependent on the peak rate-of-change of the stroke current. The resulting voltage waveform is determined by the time-derivative of the stroke current.

In general, induced voltage waveforms on overhead conductors (that result from lightning strokes) are a quick unipolar pulse followed by a long decaying tail. Peak currents in these loops can be theoretically bounded by considering the load to be a short circuit.

4.5.5 Potential Impact. Depending on the severity of the surge and the susceptibility of the sensitive equipment, three types of occurrences are possible (in addition to damage caused to cables and conductors): data disruption, hardware stress, and hardware destruction.

4.5.5.1 Data Disruption. Signal carrying circuits are susceptible to surge interference via conduction, inductive and capacitive coupling, and electromagnetic radiation. When surges are observed on signal lines, it is often assumed, just because the signal circuits are still working, that the noise is below the threshold, which can cause problems. This is not so [B16].

Digital circuits characteristically latch in either a "high" or a "low" state in which they are relatively stable. It takes a strong, deliberate signal to upset a latched circuit from one state to the other. Moreover, most circuits spend most of their working life in one state or the other, and very little time in transition between states.

However, when a bi-stable circuit is in transition between states, it is very susceptible to interference. The circuit behaves as a positive feedback amplifier and can amplify very weak signals to the point of saturating its switching semiconductor. Thus, even very low-magnitude surges can cause data corruption or upset if they occur at the moment of a state transition. A surge has a 50/50 chance of driving the circuit in the opposite direction to that which was intended, causing a data error by changing the digital signal from its intended "high" or "low" state. These data errors may be immediately obvious or may only be evident under a unique set of conditions that occur infrequently.

When recorded line-voltage disturbances coincide with computer malfunctions, it is often assumed that the line voltage change was responsible for the malfunction. Although this is a possibility, a more likely cause is the secondary effect of a rapid rate of change of current in ground conductors that creates surge voltages among different parts of the common ground referencing system. This problem often can be corrected by reconfiguring the ground referencing system interconnections rather than filtering the surge from its supply voltage (FIPS PUB 94 [11]).

Many sensitive electronic loads contain amplifiers that are routinely used to amplify the data signals. Any unwanted signal (i.e., noise) entering the input to such amplifiers, where that noise signal is completely or partially within the amplifier bandwidth, is amplified along with the desired signals. Once this happens, the unwanted, amplified noise signal is distributed within the system in a stronger form than when it entered.

4.5.5.2 Hardware Stress. A single lightning or switching surge often causes physical damage that contributes to latent device failures. Exposure to lower magnitude surges cause either a gradual performance deterioration and/or intermittent operation. In such cases, it is often difficult to differentiate between software and hardware induced errors. Latent failures are observed primarily in semiconductor devices and insulating materials.

4.5.5.3 Hardware Destruction. The third possible impact of surges is the total destruction of hardware components in a single incident. Table 4-6 shows the threshold voltages and energy levels for destruction of selected semiconductors that are commonly used in sensitive electronic equipment [B13], [B16]. Similarly, larger devices, such as transformers, relay coils, and power supply components, can be destroyed.

4.5.6 Surge Voltage Frequency. Knowledge of the frequency distribution of voltage (or current) within surges can be important in assessing their impact on sensitive electronic equipment. Depending on the surge waveshape, its voltage (and current) spectra, $V(\omega)$ and $I(\omega)$, can vary considerably. The propagation of current surges having high-frequency components requires paths that are of low impedance at the same high frequencies.

Table 4-6
Thresholds of Failure of Selected Semiconductors

Semiconductor Device Type	Disruption Energy (joules)	Destruction Energy (joules)
Digital integrated circuits	10^{-9}	10^{-6}
Analog integrated circuits	10^{-8}	10^{-6}
Low-noise transistors and diodes	10^{-7}	10^{-6}
High-speed transistors and ICs	10^{-6}	10^{-5}
Low-power transistors and signal diodes	10^{-5}	10^{-4}
Medium-power transistors	10^{-4}	10^{-3}
Zeners and rectifiers	10^{-3}	10^{-2}
High-power transistors	10^{-2}	10^{-1}
Power thyristors and power diodes	10^{-1}	10^{-0}

Fig 4-15 depicts the frequency spectra (Fourier Transforms) of five common, standard surge voltage waveforms (IEEE Std C62.41-1991 [8], [B20], [B44]). The 0.0 dB reference level is 1 V or 1 A. The peak voltage is 6 kV for both the 1.2/50 ms and 100 kHz Ring Wave, 4 kV for the EFT (electrical fast transient), and 0.6 kV for the 10/1000 µs surges, respectively. The peak current is 3 kA for the 8/20 ms surge. Fig 4-15 indicates that most of the commonly utilized surge spectra have relatively large voltage (current) components between dc and 100 kHz. The shorter rise-time surges (e.g., EFT) have larger fractions of their total energy content at higher frequencies.

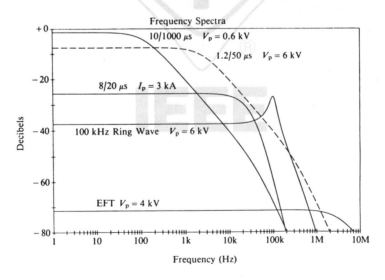

Source: [B44].

**Fig 4-15
Frequency Spectra of Common Surge Test Waveforms**

4.6 Grounding System. Sensitive electronic systems must be solidly grounded, i.e., directly connected with earth as required by either the NEC [2] or ANSI/NFPA 780-1992 [5], or both. Grounding systems designed for a sensitive electronic equipment facility can be conceptualized as having three distinct, solidly interconnected, functional subsystems. They are as follows:
  (1) Fault/personnel protection subsystem (NEC Equipment Grounding System)
  (2) Signal reference subsystem
  (3) Lightning protection subsystem

These functional subsystems are solidly interconnected to a common earth electrode subsystem.

4.6.1 Earth Electrode Subsystem. The earth electrode subsystem establishes the facility earth ground-reference for lightning, electrical fire, and shock hazard purposes only (i.e., safety purposes only). Signal transport processes and the internal signal processes of equipment are not benefitted by this system nor connections made to it except from a safety standpoint. Specific design criteria for the earth electrode subsystem are provided in the NEC [2].

4.6.2 Grounding for Fault/Personnel Protection Subsystem. This subsystem is known within the NEC [2] as the "equipment grounding system." Its primary purpose is safety. It generally has unknown characteristics regarding its impedance (versus frequency), and may be single-point, multiple-point, radial, or hybrid in some manner. It, in general, has an unknown bandwidth. It is only known to be constructed for safety reasons, and in a robust fashion per the NEC. The grounding configuration for fault/personnel protection subsystem is shown schematically in Fig 4-16.

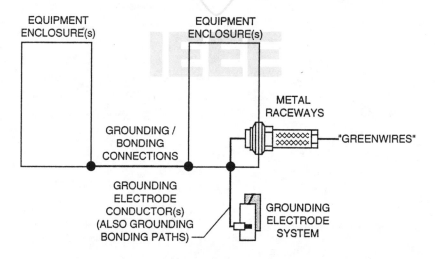

Fig 4-16
Grounding for Fault/Personnel Protection Subsystem

4.6.3 Signal Reference Subsystem. Fault/personnel protection grounding systems employing long ground conductors in facilities where high frequencies are present exhibit high impedances in the frequency ranges of interest. Therefore, they should not be used solely to provide a high-frequency reference for sensitive equipment.

Both single-point and multiple-point grounding systems, which employ long ground conductors, exhibit higher impedances at higher frequencies. Therefore, signal reference subsystems require the existence of a structure that achieves the benefits of an equipotential ground plane throughout the frequency range of interest (often from dc to tens of megahertz), as shown in Fig 4-17.

These equipotential ground plane structures achieve low impedances over large frequency ranges by providing a multitude of parallel paths among the various circuits attached to them. It is also true that for each frequency referenced within these planes, there exists at least one path that corresponds to the high-impedance quarter-wave resonance for that frequency; but this effect is negated by other paths that are half-wavelength and multiples of the fundamental resonant point. The combination of these later paths are of much lower impedance and they act as shunt-paths to the high-impedance paths. There exist an infinite number of parallel paths for current flow in the plane, the combination of these paths result in very low current densities in the plane. Low current densities throughout the plane imply equally low-voltage drops throughout the plane.

Therefore, signal reference subsystems, with equipotential ground plane structures, provide the equipotential signal-grounding means of choice, when

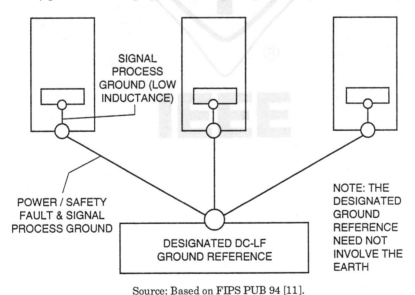

Source: Based on FIPS PUB 94 [11].

**Fig 4-17
Combined Safety and Signal Reference Grounding Subsystems**

signal frequencies range from dc to tens of megahertz. They assure that minimal voltage variances exist among the connected signal circuits and interconnected equipment.

4.6.3.1 Equipotential Plane. An equipotential ground plane is a mass (or masses) of conducting material that, when bonded together, provide a low impedance to current flow over a large range of frequencies [B37].

Advantages of an equipotential plane are as follows:
(1) Low-impedance return path for RF noise currents;
(2) Containment of EM (noise) fields between their source (cable, etc.) and the plane;
(3) Increased filtering effectiveness of contained EM fields;
(4) Shielding of adjacent sensitive circuits or equipment.

Embodiments of equipotential plane structures include the following:
(1) Conductive grid embedded in or attached to a concrete floor;
(2) Metallic screen or sheet metal under floor tile;
(3) Ceiling grid above sensitive equipment;
(4) Supporting grid of raised access flooring (computer rooms, etc.).

The concept of an equipotential reference plane can be employed within a portion of a single sensitive equipment enclosure, among various interconnected equipment, or over an entire facility. In all cases, it is bonded to both the "local building ground" and to the "grounding electrode conductor" per the NEC [2].

Within sensitive equipment cabinets, all related components, signal return leads, backplanes, etc., must be connected via short (less than 5-10% of the wavelength of the highest frequency of concern) conductors to the equipment chassis that form the equipotential plane. All similar equipment-level equipotential planes must be connected to a room-level equipotential plane via multiple (short) conductors and to the "grounding electrode conductor." The room-level equipotential plane must, in turn, be connected to one or more building-level equipotential plane(s) via multiple (short) conductors. This process continues until total sensitive electronic equipment system of interest is interconnected to one large continuous equipotential plane [B37]. The interconnecting conductors are preferred to have thin-wide cross sections to minimize their impedance at higher frequencies.

4.6.3.2 Frequency Requirements. Surges having high-frequency components require current return paths that are of low impedance at the same high frequencies. Therefore, signal reference grounding systems, which provide the required low-impedance return paths, must be designed for low-impedance characteristics over large frequency ranges, e.g., dc to tens of megahertz. Fig 4-18 shows the residual voltage versus conduction bandwidth for the IEEE Std C62.41-1991 [8] 100 kHz Ring Wave. This waveform is selected to show that such a commonly occurring surge possesses several hundred volts at frequencies greater than 1 MHz. Surge amplitudes of the order of 100 V are known to be destructive in digital circuits; therefore, the signal reference grounding system must exhibit low impedances at frequencies greater than 1–10 MHz. The upper frequency limit of practical interest (today) for most commercial equipment is considered to be in the range of 25–30 MHz. Equipotential-plane structures that possess well-behaved characteristics over this frequency range are reviewed in 4.6.3.1.

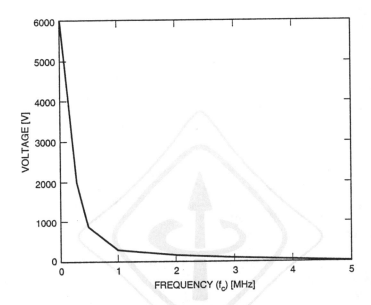

Fig 4-18
Residual Surge Voltage Versus Frequency
for 100 kHz Ring Wave

4.6.3.3 Attachment to Earth Electrode Subsystem. Equipotential plane(s) must be grounded per the NEC [2] and ANSI/NFPA 780-1992 [5]. These connections are for safety and protection from lightning surge-related sideflashes.

4.6.4 Lightning Protection Subsystem. The sole purpose of the lightning protection subsystem is the safe transport of lightning related currents through the facility to the earth grounding electrode subsystem. This is accomplished via providing highly conductive paths to direct the lightning stroke current to/from earth, while minimizing alternate paths via other items within the building. These conductors form guide paths, but do not particularly control potentials over their paths.

The lightning protection subsystem is not required by code to present any particular value or range of impedances to the lightning current that may be impressed upon it. Neither the NEC [2], ANSI/NFPA 75-1992 [3], nor ANSI/NFPA 780-1992 [5] establish impedance limits on the earth ground electrode system associated with the lightning protection system. Instead of lower resistance connections to earth, these codes favor increased frequency of bonding of the lightning conductor system to other grounded conductors within the building. This approach results in a means of reducing dangerous side-flashes, and the use of more (parallel) down-conductor paths throughout the building.

4.7 Shielding Concepts. The objective of both electromagnetic and electrostatic shielding is the significant reduction or elimination of the incidence of magnetic (or electric) fields from sensitive circuits. The basic approach is to interpose between the field source and the sensitive circuit a barrier of conducting material. Then, as changing field flux attempts to penetrate the barrier, it produces eddy currents in the barrier whose fields oppose the field of the inducing source. This allows the sensitive circuit to experience only the net-field which, depending on the barrier material and geometry, can be considerably less than the source field [B17].

Closed-form analytical solutions for several geometries are possible [B7]. Generally, it is necessary to solve the Laplace equation in the free space regions on either side of the barrier and the diffusion equation within the barrier material. These solutions are then matched at the boundaries. Several approximation techniques are also known [B45]. Specific shielding design considerations are presented in Chapter 5.

4.7.1 Electrostatic Shielding. Electrostatic shielding consists of conductive barriers, metal enclosures, or metal conduits or cable coverings around sensitive circuits. The electrostatic shield acts as a capacitive voltage divider between the field source and sensitive circuit, as shown in Fig 4-19.

The voltage divides inversely as the capacitance. Therefore, the fraction, $C_{32}/(C_{32} + C_{13})$, of the voltage between the source and sensitive circuit will appear on the shield. C_{32} is the capacitance between conductor-3 and conductor-2, and C_{13} is the capacitance between conductor-1 and conductor-3.

Thus, high-frequency grounding of the shield (per 4.6.3), and adjustments in the relative capacitance (physical separations and geometries), are the principal design factors for electrostatic shields. Shielded isolation transformers are an excellent example of the benefits of electrostatic shielding.

In order to be effective, shields must be grounded via low-impedance paths at the frequencies of interest. Long grounding conductors and long (single-grounded) shields exhibit reduced effectiveness at high frequencies, due to inductive reactance in the grounding conductor or shield. Therefore, very short grounding/bonding leads must be used, and they must be connected at the nearest equipment ground, i.e., in the example of the transformer, its metal case/enclosure. Long shields need to be grounded at multiple locations along their length. Cable shields must be either grounded at both ends or grounded at one end, and grounded via a transient surge suppression device (TVSS) at the opposite end.

FUNDAMENTALS

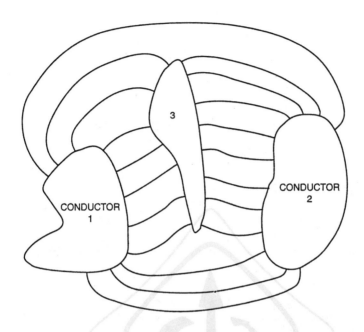

**Fig 4-19
Electrostatic Field Between Charged Conductors**

4.7.2 Electromagnetic Shielding. Effective electromagnetic shielding also consists of schemes such as high-frequency grounded conductive barriers, metal enclosures, metal conduits, and cable coverings around sensitive circuits. The objective of electromagnetic shielding is the minimization of magnetic flux coupling (mutual inductance) from a (power) source to the sensitive (e.g., control) circuit. The following generalizations are also pertinent:
  (1) Minimize the mutual inductance by physically separating the source and sensitive circuit.
  (2) Minimize the area of the sensitive circuit to reduce the number of flux lines intercepted from the source.
  (3) Use twisted pair conductors in the sensitive circuits to take advantage of the twisting wherein about half the stray magnetic flux couples into the sensitive circuit in each direction, thus giving a small net flux coupling.
  (4) Close spacing of the source (power circuit) conductors such that they appear as one conductor with equal and opposite currents, producing a minimum magnetic field.

4.8 Bonding Concepts. Effective bonding consists of a set of conductor interconnections and terminations that together, with the conductors, form a low-impedance path (at all frequencies of interest) for the flow of current through the electrical system of interest.

The objective is that each termination (bond) be such that the electrical properties of the total path are a function of the connected elements, and not the interconnections. Conversely, poor bonding is often the principal cause of many hazardous and noise-producing situations, e.g., unacceptable voltage drops, heat generation, intermittent operation, electrical noise, and high-resistance grounds.

It must be stressed that the low- and high-frequency characteristics of most bonding techniques are quite different. Their high-frequency characteristics are of particular importance for most sensitive electronic equipment applications.

The following factors are important when characterizing alternative bonding methods:

(1) *Contact resistance.* Contact resistance of conductor and shield terminations, and their aging, are of importance.
(2) *Dissimilar materials.* Dissimilar materials are problematic in that they often set up galvanic half-cells or rectifying junctions.
(3) *Skin effect.* High-frequency currents do not penetrate into high-conductivity materials. Therefore, the high-frequency impedance of bonds must be assessed.
(4) *Bond reactance.* Bond size, geometry, and the physical relationship between conductors being bonded can introduce reactive components into the impedance of the bond.

4.9 References. This standard shall be used in conjunction with the following publications. When the following standards are superseded by an approved revision, the revision shall apply:

[1] ANSI C84.1-1989, American National Standard for Electric Power Systems and Equipment—Voltage Ratings (60 Hz).[11]

[2] ANSI/NFPA 70-1993, National Electrical Code.[12]

[3] ANSI/NFPA 75-1992, Protection of Electronic Computer/Data Processing Equipment.

[4] ANSI/NFPA 77-1988, Recommended Practice on Static Electricity.

[5] ANSI/NFPA 780-1992, Lightning Protection Code.

[6] IEEE Std C57.12.00-1987, IEEE Standard General Requirements for Liquid-Immersed Distribution, Power, and Regulating Transformers (ANSI).[13]

---

[11] ANSI publications are available from the Sales Department, American National Standards Institute, 11 West 42nd Street, 13th Floor, New York, NY 10036, USA.

[12] NFPA publications are available from Publications Sales, National Fire Protection Association, 1 Batterymarch Park, P.O. Box 9101, Quincy, MA 02269-9101, USA.

[13] IEEE publications are available from the Institute of Electrical and Electronics Engineers, Service Center, 445 Hoes Lane, P.O. Box 1331, Piscataway, NJ 08855-1331, USA.

[7] IEEE Std C57.12.01-1989, IEEE Standard Requirements for Dry-Type Distribution and Power Transformers Including Those with Solid Cast and/or Resin-Encapsulated Windings.

[8] IEEE Std C62.41-1991, IEEE Recommended Practice on Surge Voltages in Low-Voltage AC Power Circuits (ANSI).

[9] IEEE Std 142-1991, IEEE Recommended Practice for Grounding of Industrial and Commercial Power Systems (IEEE Green Book).

[10] IEEE Std 446-1987, IEEE Recommended Practice for Emergency and Standby Power Systems for Industrial and Commercial Applications (IEEE Orange Book).

[11] Federal Information Processing Standards Publication 94: *Guideline on Electrical Power for ADP Installations*, Sept. 21, 1983.[14]

4.10 Bibliography

[B1] Allen, G. W. and D. Segall, "Monitoring of Computer Installations for Power Line Disturbances," *IEEE Winter Power Meeting Conference Paper*, WINPWR C74 199-6, 1974 (abstract in *IEEE Transactions on PAS*, Vol. PAS-93, Jul/Aug 1974, p. 1023).

[B2] ANSI/UL 1561-1986, Large General Purpose Transformers.

[B3] Arrillaga, J. et al., *Power System Harmonics*, New York: J. Wiley & Sons, 1985.

[B4] Blake, L. V. *Antennas*, New York: John Wiley & Sons, 1966.

[B5] Boxleitner, W. *Electrostatic Discharge and Electronic Equipment*, New York: IEEE Press, 1989.

[B6] Boyce, C. F., Ch. 25: "Protection of Telecommunication Systems," Vol. 2, *Lightning Protection*, In *Lightning*, R. H. Golde (ed), Academic Press, 1977.

[B7] Carter, G. W. *The Electromagnetic Field in Its Engineering Aspects*, London, New York, and Toronto: Longmans, Green and Co., 1954.

[B8] Cianos, N. and E. Pierce, "A Ground-Lightning Environment for Engineering Usage," Stanford Research Institute Project 1834, McDonald Douglas Astronautics Co., Contract No. L.S. 2817-A3, August, 1972.

---

[14] FIPS documents are available from the National Technical Information Service (NITS), U. S. Dept. of Commerce, 5285 Port Royal Rd., Springfield, VA 22161.

[B9] Crepaz, S., "Eddy-Current Losses in Rectifier Transformers," *IEEE Transactions on Power Apparatus and Systems*, Vol. PAS-89, No. 7, Sept/Oct 1970, pp. 165–56.

[B10] *The Dranetz Field Handbook for Power Quality Analysis*, Edison, NJ: Dranetz Technologies, Inc., 1991.

[B11] Edison Electric Institute, Power Indices, ASAI (Average Service Availability Index), Washington, DC.

[B12] Emanuel, A.E., X. Wang, "Estimation of Loss of Life of Power Transformers Supplying Nonlinear Loads," *IEEE Transactions on Power Apparatus and Systems*, Vol. PAS-104, No. 3, March 1985.

[B13] Gallace, L. and H. Pujol, "The Evaluation of CMOS Static-Charge Protection Networks and Failure Mechanisms Associated With Overstress Conditions as Related to Device Life," *Reliability Physics Symposium Proceedings*, April 1977.

[B14] Golde, R. H., Ch 17: "The Lightning Conductor," Vol. 2, "Lightning," In *Lightning Protection*, R. H. Golde (ed), London: Academic Press, 1977.

[B15] Goldstein, M. and P. D. Speranza, "The Quality of U. S. Commercial ac Power," *INTELEC* (IEEE International Telecommunications Energy Conference), 1982, pp. 28–33 [CH1818-4].

[B16] Greason, W. D. *Electrostatic Damage in Electronics: Devices and Systems*, New York: J. Wiley & Sons, 1987.

[B17] Greenwood, A. *Electrical Transients in Power Systems*, 2nd ed., New York: J. Wiley & Sons, 1991.

[B18] Gruzs, T. M., "A Survey of Neutral Currents in Three-Phase Computer Power Systems," *IEEE Transactions on Industry Applications*, Vol. IA-26, No. 4, July 1990.

[B19] Hwang, M. S., W. M. Grady, H. W. Sanders, Jr., "Assessment of Winding Losses in Transformers Due to Harmonic Currents," *Proceedings, IEEE International Conference on Harmonics in Power Systems*, Worcestor Polytechnic Institute, pp. 119–24, October 1984.

[B20] IEC Publication 801-4 (1988), Electromagnetic compatibility for industrial-process measurement and control equipment, Part 4: Electrical fast transient/burst requirements.

[B21] IEC Publication 555-1 (1982), Disturbances in supply systems caused by household appliances and similar electrical equipment, Part 1: Definitions.

[B22] IEC Publication 555-2 (1982), Disturbances in supply systems caused by household appliances and similar electrical equipment, Part 2: Harmonics.

[B23] IEC Publication 555-3 (1982), Disturbances in supply systems caused by household appliances and similar electrical equipment, Part 3: Voltage fluctuations.

[B24] IEEE Power Systems Harmonics Working Group Report, Bibliography of Power System Harmonics, Parts I & II, *IEEE Transactions on Power Apparatus and Systems*, Vol. PAS-103, No. 9, September 1984.

[B25] IEEE Std C57.110-1986, IEEE Recommended Practice for Establishing Transformer Capability When Supplying Nonsinusoidal Load Currents (ANSI).

[B26] IEEE Std C63.4-1991, Methods of Measurement of Radio-Noise Emissions from Low-Voltage Electrical and Electronic equipment in the Range of 10 kHz to 1 GHz.

[B27] IEEE Std 519-1992, IEEE Guide for Harmonic Control and Reactive Compensation of Static Power Converters (ANSI).

[B28] Keeling, M., *Evaluation and Design Reference Guide for the Grounding, Bonding, Shielding and Surge Suppression of Remote Computer Systems*, USDA Forest Service, 1987.

[B29] Key, T. S. "Diagnosing Power Quality Related Computer Problems," *IEEE Transactions on Industry Applications*, Vol. IA-15, No. 4, July/Aug 1979.

[B30] Lee, K. S. *EMP Interaction: Principles, Techniques, and Reference Data*, New York: Hemisphere Publishing Corporation, 1986.

[B31] Martzloff, F. D., "Coupling, Propagation, and Side Effects of Surges in an Industrial Building Wiring System," *IEEE Transactions on Industry Applications*, Vol. IA-26, No. 2, March/April 1990, pp. 193–203.

[B32] Martzloff, F. D., "Varistor Versus Environment: Winning the Rematch," *IEEE Transactions on Power Delivery*, Vol. PWRD-1, No. 2, April 1986, pp. 59–66.

[B33] Martzloff, F. D. and T. M. Gruzs, "Power Quality Surveys: Facts, Fictions, and Fallacies," *IEEE Transactions on Industry Applications*, Vol. 24, No. 6, Nov/Dec 1988, pp. 1005–18.

[B34] Martzloff, F. D. and T. F. Leedy, "Electrical Fast-Transient Tests: Applications and Limitations," *IEEE Transactions on Industry Applications*, Vol. IA-26, No. 1, Jan/Feb 1990, pp. 151–59.

[B35] McCann, G. D., "The Measurement of Lightning Currents in Direct Strokes," *AIEE Transactions*, Vol. 63, 1944, pp. 1157–64.

[B36] McEachern, A., *Handbook of Power Signatures*, Basic Foster City, CA: Measuring Instruments, 1988.

[B37] MIL-HNDBK-419, Grounding, Bonding, and Shielding for Electronic Equipments and Facilities, Vol. 1 (Basic Theory), Vol. 2 (Applications).

[B38] MIL-STD-188/124, Grounding, Bonding and Shielding for Common Long Haul Tactical Communication Systems Including Ground Based Communications—Electronics Facilities and Equipments.

[B39] Nordgard, J. D., and C. L., Chen, "FAA Lightning Protection Study: Lightning Induced Surges on Buried Shielded Transmission Lines; Numerical Analysis and Results," FAA Report #FAA-RD-77-83, FAA-Georgia Institute of Technology Workshop on Grounding and Lightning Technology, May 1977.

[B40] *Radio Engineers' Handbook*, New York: McGraw-Hill Book Company.

[B41] Rudenberg, *Transient Performances of Electrical Power Systems*, Boston, MA: MIT Press, 1970.

[B42] Skilling, H. H., *Electrical Engineering Circuits*, New York: J. Wiley & Sons, 1965.

[B43] Standler, R. B., "Calculation of Energy in Transient Overvoltages," *IEEE National Symposium on Electromagnetic Compatibility*, 1989, pp. 217–22.

[B44] Standler, R. B., *Protection of Electronic Circuits for Overvoltages*, New York: J. Wiley & Sons, 1989.

[B45] Stratton, J. A. *Electromagnetic Theory*, New York: McGraw-Hill Book Co., 1941.

[B46] Sunde, E. D., "Earth Conduction Effects on Transmission Systems," Van Nostrand Company, 1949, and Dover Publications, 1968.

[B47] Sunde, E. D., "Lightning Protection for Buried Toll Cable," *Bell System Technical Journal*, No. 24, April 1945.

[B48] UL 1562-1988, Transformers, Distributers, Dry-Type—Over 600 Volts.

[B49] Van Keuren, E. "Effects of EMP Induced Transients on Integrated Circuits," *IEEE Symposium on Electromagnetic Compatibility*, 1975, pp. 1-5.

[B50] Vorgucic, A. D., "Condition of Evaluation of the Protection Zone of the Lightning Rod," FAA Report #FAA-RD-78-83, FAA-Georgia Institute of Technology Workshop on Grounding and Lightning Technology, 1978.

[B51] Waters, W. *Electrical Induction from Distant Current Surges*, Englewood Cliffs, NJ: Prentice-Hall, Inc., 1983.

Chapter 5
Instrumentation

5.1 General Discussion. Power quality site surveys and longer-term monitoring programs both require proper instrumentation in order to be effective. There is a wide variety of measuring equipment available for use.[15] The challenge is in selecting the most appropriate instrumentation for a given test or measurement [B5].[16]

5.2 Wiring and Grounding Measurement Instruments. Problems in industrial/commercial premises wiring and grounding account for a large share of all reported power-quality problems. The greatest number of wiring and grounding problems is in feeders and particularly branch circuits serving the critical load.

Chapter 6 describes the recommended practice for conducting a survey of the power distribution system, including the following steps:

(1) Conduct a wiring verification of phase voltages, currents, phase rotation, load balance, equipment grounding conductor impedance, neutral impedance, and the presence of the required neutral-ground bond at the service equipment and at downstream transformer secondaries.
(2) Test for wiring errors at the panel or outlet of interest. Missing connections, including open equipment grounding conductor, open neutral, or open phase conductors.
(3) Test for improper connections, including reversed phase/neutral or neutral/equipment grounding conductors.
(4) Poor quality connections, identified through continuity, voltage, and impedance measurements.
(5) Visually inspect, or measure for, adequate rms current-carrying capability of neutral and phase conductors.

Recommended instruments for conducting these measurements are shown in Table 5-1. These instruments are discussed further in the following sections.

CAUTION! **Workers involved in opening energized power panels are required to abide by the prescriptions of ANSI/NFPA 70E-1988 [B2] concerning appropriate protective equipment, as well as government regulations codified in OSHA CFR 1910 [B3] and 1926 [B4], and in ANSI C2-1990 [B1].**

---

[15] Figures 5-1 through 5-10 represent examples of different types of measuring instruments discussed in this chapter. This Recommended Practice does not intend to imply that the manufacturers who contributed photographs for these figures make the only, or the preferred, instrument of this type.

[16] The numbers in brackets correspond to those of the references in 5.19; when preceded by the letter "B," they correspond to those in the bibliography in 5.20.

Table 5-1
Recommended Test Instruments for Conducting a Site Survey

Instrument	Minimum Required Instrumentation			Multiple Function or Special Purpose Instrumentation				
	True RMS Multimeter	True RMS Clamp-on Ammeter	Ground Impedance Tester	Earth Ground Tester	Oscilloscope With Current Transducer	Oscilloscope With Line Decoupler	Power Disturbance Monitor	Spectrum Analyzer
Measurement	Voltage, Continuity	Current	Impedance	Resistance, Impedance	Current, Waveforms	Voltage, Waveforms	Voltage, Current, Waveforms, Harmonics	Harmonics, Noise, Spectra
Neutral-Ground Bond								
(1) Grounding electrode conductor connections	*							
(2) Main bonding jumper connections	*							
(3) Extraneous bonds downstream from service entrance and/or separately derived secondary bond	•							
Neutral Conductor Sizing, Routing								
(1) Parity or greater than phase conductor neutral sizing		•						
(2) Shared (daisy-chained) neutrals		•			•	•	•	
Grounding-Grounding Electrode System								
(1) Equipment grounding conductor impedance			•					
(2) Equipment grounding conductor integrity when used with supplementary grounding electrodes	•							
Dedicated Feeders, Direct Path Routing								
(1) Other equipment on the circuit of interest					•		•	
(2) Equipment grounding conductor impedance			•					
(3) Mixed grounding means problems					•			
Grounding Electrode Impedance								
(1) Resistance of the grounding electrode				•				
(2) Grounding electrode conductor integrity		•						

* Micro-ohmmeter

Table 5-1 (continued)
Recommended Test Instruments for Conducting a Site Survey

Instrument	Minimum Required Instrumentation				Multiple Function or Special Purpose Instrumentation			
	True RMS Multimeter	True RMS Clamp-on Ammeter	Ground Impedance Tester	Earth Ground Tester	Oscilloscope With Current Transducer	Oscilloscope With Line Decoupler	Power Disturbance Monitor	Spectrum Analyzer
Measurement	Voltage, Continuity	Current	Impedance	Resistance, Impedance	Current, Waveforms	Voltage, Waveforms	Voltage, Current, Waveforms, Harmonics	Harmonics, Noise, Spectra
Continuity of Conduit/Enclosure Grounds								
(1) Metallic enclosure, conduit, raceway, panel board continuity	•							
(2) Bonding jumpers where nonmetallic conduit is used	•		•			•	•	
(3) Continuity of expansion joints, telescoping raceway, and wiremolds	•		•			•	•	
Separately Derived System Grounding								
(1) Verify neutral as separately derived and not interconnected	•		•					
(2) Impedance of neutral-ground bond on secondary			•			•	•	
(3) Grounding electrode conductor connections			•			•	•	
Isolated Ground Systems								
(1) Conductor insulation from conduit ground system	•							
Power Disturbances								
(1) Under- or overvoltages	•					•	•	
(2) Momentary sags and swells						•	•	
(3) Surges						•	•	•
(4) Notches						•	•	
(5) Outages and momentary outages						•	•	
(6) Electrical noise							•	•
(7) Harmonics							•	•
(8) Frequency deviations							•	•

5.3 Infrared Detector. The overheating of transformers, circuit breakers, and other electrical apparatus is often impossible to detect from current and voltage measurements. Infrared detectors produce images of the area under investigation. Overheated areas become apparent in contrast to normal temperature images.

5.4 Root-Mean-Square (RMS) Voltmeters. Alternating-current (AC) voltmeters are designed to measure the effective value, or equivalent heating value, of the ac voltage. The voltmeter converts the ac being measured either directly to heat, with the thermocouple-type voltmeter; or to an equivalent dc voltage. This equivalent value is usually expressed as the root-mean-square (rms) value of the voltage.

AC voltmeter may be true-rms responding, quasi-rms responding, average responding, or peak responding. All rms meters are calibrated so as to read in rms units. Any of these techniques will work equally well as long as the ac being measured has a purely sinusoidal waveform. Large errors result when the waveforms include significant distortion or noise.

AC voltmeters that respond to average, peak, or rms values are commonplace. Typical voltmeter are an "average actuated, rms calibrated" device. The assumption is that the measured wave is sinusoidal, and that the ratio between the rms and average values is always a constant. There can be significant errors if the waveforms are not sinusoidal. Recommended practice is to use true rms voltmeters when the shape of the waveform is unknown or other than a pure sine wave.

5.5 True RMS Voltmeters. True rms reading voltmeters indicate the square root of the sum of the squares of all instantaneous values of the cyclical voltage waveform.

Normal voltage wave shapes in ac power distribution systems are designed to be predominantly sinusoidal. The heaviest electrical loads have historically been resistance elements used for heating and lighting, and induction motor loads. These represent "linear" loads in which current is approximately proportional to voltage, and predominantly sinusoidal.

Proliferation of power electronic loads have changed the nature of the loads that might be expected on ac power distribution systems. They represent nonlinear loads characterized by distorted voltage and current waveforms. Large peak currents are drawn at the peak of the voltage waveform but not at other times. The result is a distorted load current wave and a distorted voltage drop in the ac power source impedance. Use of true rms instruments for voltage and current measurements is, therefore, the recommended practice. A variety of true rms voltmeters are in use, including the thermocouple type, square-law type, and sampling type.

5.5.1 Thermocouple Type. The rms value of a voltage is defined in terms of the heat it will produce in a resistive load. Thus, a natural way to measure true rms voltage is by means of a thermocouple device, which includes a heating element and a thermocouple in an evacuated chamber. The heating element produces heat in proportion to the rms voltage across it, and the thermocouple produces a dc voltage in proportion to the generated heat. Since thermocouples are affected by inherent nonlinearities and by environmental temperature, a second thermocouple is typically added in a feedback loop to cancel these effects and produce a workable rms-responding voltmeter.

5.5.2 Square-Law Type. Another approach to true rms measurements is to use the nonlinear characteristics of a P-N junction to produce an analog squaring circuit. Once a squarer is available, the rms voltage is calculated.

5.5.3 Sampling Devices. Yet another approach to true rms measurement is to digitally sample the incoming ac voltage at relatively high rates, square the sampled values, average the squared samples over one or more complete ac cycles, and take the square root of the result. This technique lends itself nicely to digital manipulation without the drifting overtime and temperature inherent in analog square-law devices.

5.6 Direct-Reading Ammeters. Direct-reading ammeters are those employed with a series shunt and that carry some of the line current through them for measurement purposes. They are part of the circuit being measured.

Direct-reading ammeters include electrodynamometer types, moving-iron-vane meters, and thermocouple types that drive dc-responding d'Arsonval meters. All of these ammeter types respond directly to the current squared, and are not true rms meters.

Use of true rms ammeters is recommended practice due to nonlinear load currents.

5.7 True RMS Ammeters. True rms ammeters include two types of indirect reading ammeters: current transformers and Hall-effect types. The voltage meter considerations also apply to current meters. For unknown or nonsinusoidal currents, recommended practice is to use true rms ammeters. The growing number of power electronic loads has increased the likelihood of current waveform distortion. The clamp-on type of true rms ammeter (Fig 5-1) is recommended due to ease of use, although other types would be satisfactory.

5.7.1 Current-Transformer (CT) Ammeters. A transformer is commonly used to convert the current being measured to a proportionately smaller current for measurement by an ac ammeter. There is very little resistive loading with these ammeters, and when a split-core transformer is used, the circuit is not interrupted. These ammeters may not be true rms reading meters.

The transformer inductively couples the current being measured to a secondary consisting of N turns of wire (N_s). If the current being measured is I, and if we assume the primary is equivalent to a single turn, the secondary current I_s is calculated as follows:

$I_s = I/N_s$

CAUTION! Workers involved in opening energized power panels are required to abide by the prescriptions of ANSI/NFPA 70E-1988 [B2] concerning appropriate protective equipment, as well as government regulations codified in OSHA CFR 1910 [B3] and 1926 [B4], and in ANSI C2-1990 [B1].

Courtesy of John Fluke Manufacturing Co., Inc.

**Fig 5-1
Multimeters With Clamp-on Current Probes**

5.7.2 Hall-Effect Ammeters. The Hall effect is the ability of semiconductor material to generate a voltage proportional to current passed through the semiconductor, in the presence of a magnetic field. This is a "three-dimensional" effect, with the current flowing along the x-axis, the magnetic field along the y-axis, and the voltage along the z-axis. The generated voltage is polarized so that the polarity of the current can be determined. Both ac and dc currents can be measured.

Negative-feedback technology has eliminated or greatly reduced the effects of temperature variations and high-frequency noise on Hall-effect current probes. Hall-effect ammeters are affected by temperature variations (as is any semiconductor device) and by extreme high-frequency noise. Filtering is added to reduce this effect.

5.8 Current Measurement Considerations

5.8.1 DC Component on AC Current. All the ac ammeters discussed here are capable of responding to ac currents with dc components. The low-frequency response of CT-type ammeters falls off rapidly as the dc component of the measured current increases. Another possible effect of dc current arises from the fact that any magnetic core can become magnetized by passing relatively large dc currents through it. The result is a need for periodic degaussing.

5.8.2 Steady-state Values. Most multimeters commonly used by the electrical industry are intended for providing steady-state values of current or voltage. The measured rms current or voltage is sampled or "averaged" over several cycles. By necessity, real-time meters cannot display cycle-by-cycle activity for a 60 Hz system. The response time of analog meter movements is much greater than the 16 ms period of 60 Hz. In fact, digital meters deliberately delay updating the display to eliminate bothersome flicker that occurs with updates quicker than about 0.1 s.

Steady-state load current in all phases and neutral conductors should be measured with a true rms ammeter as per the wiring and grounding tests described in Chapter 6. Steady-state peak current should be measured with an oscilloscope and current probe or power monitor. Measurements with a moving coil or "peak hold" ammeter can give erroneous information.

5.8.3 Inrush and Start-up Current Values. It is often desirable to measure accurately the transient currents and voltages that result from the turn-on of electronic loads and other equipment. Where the load is largely reactive (motor, transformer), these initial currents can be several times the steady-state values.

To measure such brief currents, a fast-responding ammeter is required, along with a matching circuit to either display (as on a meter) the peak current, or record it (as on a paper graph). It is also possible to use an oscilloscope or power monitor with a fast responding CT-type current probe.

Direct-reading ammeters are far too slow to respond to rapid changes. Both the CT-type and Hall-effect ammeters are capable of response up to hundreds of megahertz, or even gigahertz, although additional circuitry must be added to hold the desired peak values.

5.8.4 Crest Factor. The ratio of peak-to-rms current is known as crest factor. This measurement is important in the assessment of nonlinear loads.

5.9 Receptacle Circuit Testers. Receptacle circuit testers are devices that use a pattern of lights to indicate wiring errors in receptacles. These devices have some limitations. They may indicate incorrect wiring, but cannot be relied upon to indicate correct wiring.

5.10 Ground Circuit Impedance Testers. Ground impedance testers (Fig 5-2) are multifunctional instruments designed to detect certain types of wiring

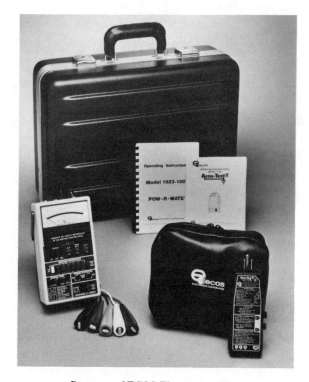

Courtesy of ECOS Electronics Corp.

**Fig 5-2
Ground Circuit Impedance Testers**

and grounding problems in low-voltage power distribution systems. Some instruments are designed for use on 120 Vac single-phase systems while others can be used on both single- and three-phase systems up to 600 Vac. The primary test function is impedance measurement of the equipment grounding conductor or neutral (grounded conductor) from the point of test back to the source neutral-ground bond. Additional test functions include detection of wiring errors (e.g., reversed polarity, open equipment grounding conductor, and open neutral), voltage measurement, the presence of neutral-ground shorts, and isolated ground shorts.

5.11 Earth Ground Resistance Testers. Ground resistance tests should be conducted with a fall-of-potential method instrument. Clamp-on instruments that do not require the grounding electrode to be isolated from the building grounding system for the test are of questionable validity for industrial and commercial buildings (IEEE Std 81-1983 [B6]). In practice, the resistance of the earth

INSTRUMENTATION

grounding electrode is tested when the building is inspected, following its construction, but at no other time.

5.12 Oscilloscope Measurements. In its simplest form, the oscilloscope is a device that provides a visual representation of a voltage plotted as a function of time. Even a limited-feature oscilloscope can be quite useful in detecting the presence of harmonics on an electrical system.

5.12.1 Line Decoupler and Voltage Measurements. Voltage measurements are relatively straightforward using an oscilloscope (Fig 5-3). The input is connected to the voltage of interest with the appropriate lead. If a voltage above the range of the oscilloscope is to be examined, probes with resistance-divider networks are available to extend the range of the instrument by a factor of 10 or

Courtesy of Oneac Corp.

Fig 5-3
Voltage Measurements Can Be Made With Oscilloscopes Using Line Attenuators

more, or capacitively coupled voltage step-down devices are available. The frequency responses of the capacitively coupled voltage step-down devices are nearly constant from the power frequency to the lower radio frequency range.

5.12.2 Clamp-on Current Transducer and Current Measurements. The oscilloscope cannot measure current directly, only a voltage produced as a current passes through a resistance. Measurements of currents based on the use of a shunt (current-viewing resistor) can be made with a differential input provided on oscilloscopes. If only a single-ended input is available, the signal is then applied between the high input and the oscilloscope chassis, creating a ground loop. Attempts are then sometimes made to break this ground loop by disconnecting the equipment safety grounding conductor (greenwire) of the oscilloscope. This practice "floating scope" is a safety risk and is prohibited (IEEE Std C62.45-1987 [1]).

Clamp-on current transformers provide a means of isolating the oscilloscope from the circuit being tested. Some models have a resistance in place across the secondary of the current transformer to facilitate use with test equipment. In cases where the user must supply the secondary resistor, the resistance should be kept to a minimum to prevent saturation of the current transformer core. If the core becomes saturated, the oscilloscope waveform will show a different harmonic content than is present in the primary circuit.

One bothersome characteristic of current transformers in general is a nonlinear frequency response. Typical current transformers give accurate current reproduction only over the range of 50 Hz to 3 kHz. Units with "flat" frequency response up through several kilohertz are available but costly. In some current probes, digital correction of frequency response is possible.

5.13 Power Disturbance Monitors. Power monitors are a new class of instrumentation developed specifically for the analysis of voltage and current measurements. At present, there are no standards for categorizing types of power disturbances recorded by these power monitors. Consequently, the type of disturbance recorded by different power monitors may vary from manufacturer to manufacturer.

Time-domain and limited frequency domain measurements are possible. Where their cost can be justified, power monitors are recommended instruments for conducting site surveys or longer-term monitoring programs. Table 5-1 lists the measurements monitors can make. It is a matter of user preference as to whether power monitors should be employed in the early stages of a site survey that are likely to concentrate on wiring and grounding measurements. The multiple-featured power monitors often contain true rms voltage and current measurement capability, which is necessary for most of these measurements.

Although developed for the common application of detecting voltage aberrations that affect the operation of electronic equipment, power disturbance monitors have many different characteristics. The differences include channel capacity, measurement performance, data output, and ancillary features that are of considerable importance to the user. Power disturbance monitors are of three basic types: event indicators, text monitors, and waveform analyzers.

5.13.1 Event Indicators. The simplest and least expensive types of power-disturbance monitors are known as event indicators. Event indicators detect, classify, and indicate power line disturbance events when they occur. Individual events are not identified by time of occurrence. Data output consists of an illuminated display or alarm that indicates the prior occurrence of a disturbance event. Event indicators are recommended for identifying the need for additional power line disturbance analysis using more sophisticated instrumentation.

5.13.1.1 Data Capture Techniques. Event indicators capture disturbance data by comparing the monitored parameter, usually ac voltage, to one or more threshold parameters. When the threshold parameter is exceeded, a disturbance event is detected and indicated. The comparison of monitored parameter to threshold parameter may be accomplished by analog techniques, digital techniques, or by combinations of analog and digital comparison circuits. Threshold parameters may be fixed or adjustable by the user over a specified range to accommodate different monitoring circumstances. Some examples of common threshold parameters include the following:
(1) *AC rms voltage.* With rms sensing or average sensing, the measurement interval should be an integral number of half cycles of the fundamental power frequency. With peak sensing, the measurement interval should be one-half cycle of the fundamental power frequency.
(2) *Surge voltage.* Peak detection should be used for disturbance events of short duration.
(3) *Frequency.* The measurement interval should be small in comparison with the duration of the disturbance to be measured.

Characteristics of threshold parameters determine the types of disturbance events that are detected. Therefore, a complete understanding of the threshold parameters of a given instrument is essential for proper usage of the event indicator.

5.13.1.2 Recording and Reporting Mechanisms. Having detected the disturbance event, event indicators store the data as a count, an amplitude, or both. Event data are then reported as a cumulative count or as an amplitude, possibly accompanied by blinking lights, audible alarms, or other forms of annunciation.

5.13.1.3 Analysis Functions. Event indicators provide minimal analytical capability. The user is alerted to the prior occurrence of a disturbance event, but lacking descriptive information and time of occurrence of individual events, the user is unable to analyze causes or consequences of the events that occurred. Therefore, very little guidance concerning the nature and solution of the suspected ac power problem is possible.

5.13.2 Text Monitors. Text monitors are power disturbance monitors that detect, classify, and record power line disturbance events (Figs 5-4 to 5-6). Individual events are recorded by time of occurrence and alphanumeric descriptions that are representative of disturbance events occurring during a given time interval. Data output may be reported on paper or electronic media, possibly accompanied by alarm annunciation.

5.13.2.1 Data Capture Techniques. Text monitors use threshold comparison techniques, which are similar to those of event indicators that were previously discussed, to detect disturbance events. Monitored parameters are continually compared to one or more threshold parameters. When a threshold parameter is exceeded, a disturbance event is detected and numerous measurements of disturbance characteristics may be stored. As with event indicators, threshold comparison may be analog or digital; fixed or adjustable over a specified range. Some examples of common threshold parameters are as follows:
  (1) *AC rms voltage.* With rms sensing or average sensing, the measurement interval should be one period or more of the fundamental power frequency. With peak sensing, the measurement interval should be no more than one-half period of the fundamental power frequency.
  (2) *Surge voltage.* Peak detection should be used for disturbance events having short duration.
  (3) *Frequency.* The measurement interval can be less frequent than that for impulses but should still be small with respect to the rms change being measured.
  (4) *Notch.* Peak detection should be used for notches having short duration.

Characteristics of the threshold parameters determine the types of disturbance events that are detected. Therefore, a complete understanding of the threshold parameters and detection methods of a given instrument is essential for proper usage of the text monitor.

5.13.2.2 Recording and Reporting Mechanisms. The recording and reporting mechanisms of text monitors facilitates the incorporation of numerous measurement capabilities. When a disturbance event is detected, these measurements are recorded to comprise an alphanumeric description that is representative of the disturbance event. The accuracy of this alphanumeric representation depends upon measurement parameters, measurement techniques, and the extent of recorded detail. An extensive variety of measurements are possible, but the most common include the following:
  (1) *Time of occurrence.* The time that the disturbance event begins should be measured with as much precision as may be required for a given application. Specifications range from the nearest second to the nearest 1 ms.
  (2) *AC rms voltage.* Each half-period of the fundamental power should be measured.

Courtesy of Monistar Systems, Inc.

**Fig 5-4
Text Monitors Can Be Used to Detect AC RMS Voltage Variations**

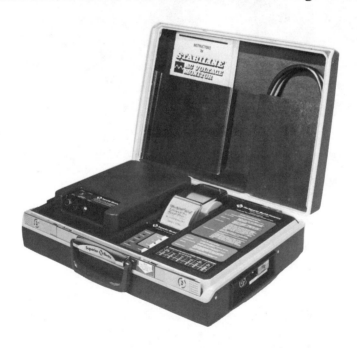

Courtesy of Superior Electric.

**Fig 5-5
Text Monitors Are Typically Lightweight and Well-Suited to Use in the Field**

CAUTION! **Workers involved in opening energized power panels are required to abide by the prescriptions of ANSI/NFPA 70E-1988 [B2] concerning appropriate protective equipment, as well as government regulations codified in OSHA CFR 1910 [B3] and 1926 [B4], and in ANSI C2-1990 [B1].**

Courtesy of Angus Electronics Co.

**Fig 5-6
The Newest Category of Text Monitors is Capable of Making Simultaneous Voltage and Current Measurements**

(3) *Surge voltage.* Peak voltage amplitude measured with respect to the power frequency sine wave. Duration, rise time, phase, polarity, and oscillation frequency may also be measured.

(4) *Frequency.* The measurement interval should be from 0.1 to 1 s.

(5) *Total harmonic distortion.* The measurement interval should be from 0.1 to 1 s. Amplitude and phase of individual harmonic numbers may also be measured.

INSTRUMENTATION

The text monitor stores all measurements of the disturbance event and then composes the measured data into an alphanumeric format that is representative of the original disturbance. A sequential series of alphanumeric descriptions is then reported to paper printout or electronic media.

Text monitors may have other features, beyond the five most common. Examples include common-mode noise detection, temperature, humidity, and dc voltage and current measurement.

5.13.2.3 Analysis Functions. The sequential recording of disturbance events, with precise time of occurrence, by text monitors enables the user to correlate specific power line disturbances with misoperation or damage of susceptible equipment. Furthermore, the alphanumeric description of the disturbance event is useful in determining the cause and probable consequences of the disturbance event. Other data contained within the alphanumeric description can be statistically related to determine the probability of various disturbances occurring at the monitored site. Analysis functions are limited only by the extent of the alphanumeric description and by the skill and experience of the user. Therefore, the analysis capabilities of text monitors may be very extensive.

5.13.3 Waveform Analyzers. Waveform analyzers are power disturbance monitors that detect, capture, and record power line disturbance events as complete waveforms supplemented by alphanumeric descriptions common to text monitors (Figs 5-7 to 5-9). The ability to capture, store, and report disturbance waveforms makes the waveform analyzer the preferred choice for intensive analysis of ac power quality. Individual disturbance events are recorded by time of occurrence with waveforms and alphanumeric measurements that are representative of disturbance events occurring during a given time interval. Data output may be reported on paper or electronic media, possibly accompanied by alarm annunciation.

5.13.3.1 Data Capture Techniques. Waveform analyzers use sampling techniques to decompose the ac voltage waveform into a series of discrete steps that can be digitally processed, stored, and eventually recombined to represent the original ac voltage waveform. Waveform sampling occurs continuously at a fixed or variable rate. High sampling rates result in better representation of the disturbance waveform and greater storage requirements.

Although waveform sampling is continuous, waveform analyzers store only the sampled data when a disturbance is detected. Disturbance detection is determined by comparison of threshold parameters with the monitored parameter. As with text monitors, threshold comparison may be analog or digital, fixed or adjustable, over a specified range.

Due to the continuous waveform sampling, threshold comparison algorithms tend to be more complex than those of text monitors. However, this complexity provides tremendous flexibility in controlling the types of disturbance waveforms that are detected. As with all power disturbance monitors, a

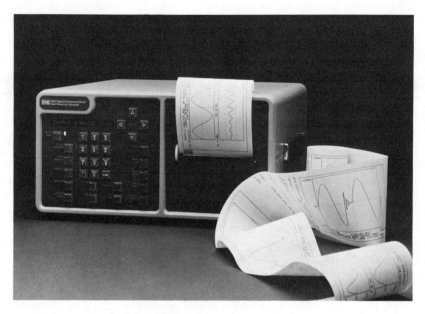

Courtesy of Basic Measuring Instruments.

**Fig 5-7
Waveform Analyzers Detect, Capture, and Record Power Line Disturbance Events as Complete Waveforms**

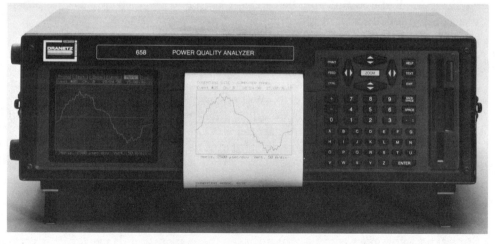

Courtesy of Dranetz Technologies, Inc.

**Fig 5-8
Data Output From Waveform Analyzers**

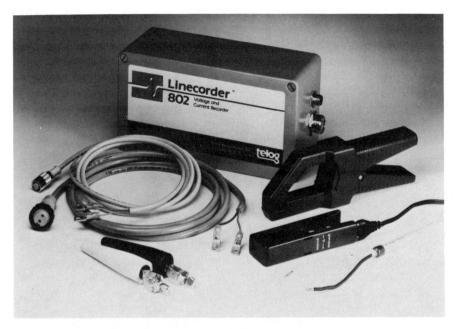

Courtesy of Telog Instruments, Inc.

**Fig 5-9
Some Waveform Analyzers are Capable of Down-loading Data
to a Personal Computer**

complete understanding of the threshold parameters and detection methods of a given instrument is essential for proper usage of the waveform analyzer.

5.13.3.2 Recording and Reporting Mechanisms. When a disturbance event is detected, the digitized samples are stored in memory. As subsequent processing, measurement, and reporting of the disturbance event will be based entirely upon the stored samples, the waveform analyzer must retain sufficient data from before and after the detection point to accurately reconstruct the entire disturbance event.

Having captured and stored the digitized data, the waveform analyzer is able to compute numerous measurements of the disturbance event. These measurements of disturbance characteristics are at least as extensive and as accurate as those available from text monitors. Furthermore, the digitized data can be formatted to provide a detailed graphic representation of the disturbance waveform.

This graphic reporting may be accomplished by paper printout or electronic media such as magnetic tape, diskettes, cathode ray tube (CRT) displays, etc. With accuracy of the graphic and alphanumeric representation of the disturbance event limited only by measurement techniques and storage capacity, waveform analyzers can provide the most complete description of a power line disturbance that is practical from a power disturbance monitor.

5.13.3.3 Analysis Functions. The graphic reporting of the disturbance waveform enables the user to perform several additional analysis functions. First, the time-based correlation of disturbance waveforms with misoperation of sensitive electronic equipment can facilitate more meaningful susceptibility testing followed by corrective design improvements. These design improvements, both at the system and equipment levels, can lead to improved immunity against ac power line disturbances. Second, the characteristic disturbance waveform of certain disturbance sources can facilitate the identification, location, and isolation of these disturbance sources. These analytical functions make the waveform analyzer most suitable for analyzing complex power quality problems when properly applied by the knowledgeable user.

5.14 Spectrum Analyzers and Computer-Based Harmonic Analysis. Harmonics, electrical noise, and frequency deviations can be measured using power monitors equipped with the appropriate measurement capabilities. These can also be measured with special-purpose harmonic meters, low-frequency or broad-band spectrum analyzers (depending on what is being measured—harmonics or noise), or with a combination of instrumentation and computer-based analysis. All instrument choices must use voltage and current probes, the considerations for which have been discussed previously with respect to oscilloscopes.

If power flow is of interest, harmonic voltage and current measurements must be made simultaneously. Harmonic analyzers and computer-based measurements use fast Fourier transforms to provide data on the amplitude, phase position, and relative contribution to the total distortion for each harmonic component.

5.15 Expert Systems. Knowledge-based and expert-system software is available for recording and analyzing power quality site survey data and reporting the results (Fig 5-10).

5.15.1 Data Collection Techniques. Expert systems use data input by the user, data encoded as procedures or as rules, and possibly data from instrumentation. Embedded and other instrumentation-based expert systems have data capture (of collected data) mechanisms that are specific to the instrument being used. Instrument-independent expert systems collect data by presenting questions to the user for response. Both instrumentation-based and instrument-independent expert systems use data encoded in the form of knowledge structures to process measurement or input data.

5.15.2 Recording and Reporting Mechanisms. Measurements and user-input data are typically recorded onto mass storage media. Communications interfaces may be used to accomplish data recording. A common technique in data recording is to store the data in an electronic data base that can be

Courtesy of Collective Intelligence, Inc.

**Fig 5-10
Expert Systems Use Data Input by the User, Software Procedures, and Rules**

accessed by the expert system. Processed data and analysis results are reported on the computer screen or by means of printed reports. Reports typically include tutorial information explaining the expert system's reasoning.

5.15.3 Analysis Functions. Expert systems for power quality analysis differ in scope and depth; hence, in analysis capabilities. Embedded and instrument-based expert systems are designed to assist in the analysis of specific measured data, including one or more types of power disturbance. Instrument-independent expert systems that are not instrument-dependent have broader scope, but perhaps less depth relative to analyzing measured data. Site survey analysis software is an example of this type of expert system, the scope of which includes wiring, grounding, surge protection, power monitoring, data analysis, and power conditioning equipment recommendation.

Expert systems can provide consistency and expert tutorial in the collection, analysis, and reporting of power quality data if appropriately applied by the user.

5.16 Electrostatic Discharge. Electrostatic charge can be measured with special hand-held meters designed for that purpose.

5.17 Radio Frequency Interference (RFI) and Electromagnetic Interference (EMI). Electric and magnetic field probes measure broadband field strength. A field-strength meter equipped with a suitable probe for electric or magnetic field sensing can be used to assess RFI or EMI more generally.

5.18 Temperature and Relative Humidity. Measure temperature and relative humidity with a power monitoring equipped with special probes. The rate of change of these parameters is at least as important as the absolute values of the temperature and relative humidity.

5.19 References. This standard shall be used in conjunction with the following publication. When the following standard is superseded by an approved revision, the revision shall apply:

[1] IEEE Std C62.45-1987, IEEE Guide for Surge Testing for Equipment Connected to Low-Voltage AC Power Circuits (ANSI).[17]

5.20 Bibliography

[B1] ANSI C2-1990, National Electrical Safety Code.

[B2] ANSI/NFPA 70E-1988, Standard for Electrical Safety Requirements for Employee Workplaces.

[B3] CFR (Code of Federal Regulations), Title 29, Part 1910: Occupational Safety and Health Standards (OSHA).

[B4] CFR (Code of Federal Regulations), Title 29, Part 1926: Safety and Health Regulations for Construction (OSHA).

[B5] Clemmensen, Jane M., "Power Quality Site Survey Instrumentation and Measurement Techniques," IEEE I&CPS (1990), Paper No. 90CH2828-2/90/000-0126.

[B6] IEEE Std 81-1983, IEEE Guide for Measuring Earth Resistivity, Ground Impedance, and Earth Surface Potentials of a Ground System.

---

[17] IEEE publications are available from the Institute of Electrical and Electronics Engineers, Service Center, 445 Hoes Lane, P.O. Box 1331, Piscataway, NJ 08855-1331, USA.

Chapter 6
Site Surveys and Site Power Analyses

6.1 Introduction. High-speed electronic systems and equipment may be more sensitive to disturbances in the ac power system than are conventional loads. The effects of power disturbances on sensitive electronic equipment can take a wide variety of forms including data errors, system halts, memory or program loss, and equipment damage. In many cases it is difficult to determine whether the system hardware and software malfunctions are actually caused by disturbances in the power system supplying the equipment. Usually, some level of survey and analysis of the power system is required to determine if power disturbances are affecting system performance.

6.2 Objectives and Approaches. The basic objectives of surveys and site power analyses are as follows:
 (1) Determine the soundness of the power distribution (wiring) and grounding system supplying the equipment.
 (2) Determine the quality of the ac voltage supplying the equipment.
 (3) Determine the sources and impact of power system disturbances on equipment performance.
 (4) Analyze the survey data to identify cost-effective improvements or corrections, both immediate and future.

It is important to keep these approaches in mind when a site is experiencing problems that appear to be power-related. All too often, corrective action (in the form of some type of power conditioning equipment) is installed in a "shotgun" attempt to solve the problem. Although this method will sometimes minimize the problem, in the majority of cases it may do little or nothing to solve the problem and can even aggravate conditions resulting in further degradation of system performance levels.

To successfully resolve problems in the power system supplying sensitive electronic equipment, a thorough analysis of the power system and loads should be conducted to define the areas of concern as accurately as possible before attempting to solve the problem. This approach can enable cost-effective solutions to be implemented that not only correct the existing conditions but also minimize future problems.

The key is to understand and define the problem before attempting to solve it. Some of the parameters that need to be defined are given here:
 (1) What sensitive electronic equipment is experiencing problems (e.g., type, location).
 (2) The types of equipment malfunctions or failures (e.g., data loss, lockups, component damage).

(3) When do the problems occur (e.g., time of day, day of week, particular system operation).
(4) Coincident problems occurring at the same time (e.g., lights flicker, motors slow down).
(5) Possible problem sources at site (e.g., arc welders, air conditioning, copy machines).
(6) Existing protection for equipment (e.g., transient voltage surge suppressor, isolation transformer).

This should provide information for a preliminary analysis to decide if immediate recommendations for remedial action can be taken.

6.3 Coordinating Involved Parties. It is the responsibility of the end user or electronic equipment owner to provide and maintain a proper supply of ac power from the utility service entrance to the equipment. In addition to the owner, other involved parties should be informed as to the objectives of the site survey. Effective communications will help assure that the recommendations for improvement or correction will be implemented in a mutually acceptable manner.

6.3.1 The Equipment User, Owner, or Customer. The user of electronic equipment is primarily concerned with the productivity of the equipment. Downtime translates into loss of production, increased operating costs, and decreased revenues and profits. Technical details on power disturbances are normally of little interest to the end user who cares only that the equipment is not performing as intended and it is costing the company money. It is important that the end user understand it is in his best interest to provide and maintain a sound power source to operate the equipment. Keeping an accurate log of equipment errors and malfunctions can provide valuable information in solving site power problems.

6.3.2 The Electronic Equipment Manufacturer/Supplier. Initially, it is the responsibility of the equipment manufacturer or supplier to provide the power, grounding, and environmental specifications and requirements for their equipment. If this has not been done, the effectiveness of the service representative will be reduced when a power-related problem develops since it is the service engineer who normally determines the problem and relates this information to the end user. When the problem areas have been defined, the recommended methods of correction should be clearly communicated to the end user so that an incorrect or partial solution does not occur.

6.3.3 The Independent Consultant. In many cases, a practical approach is to enlist the services of an independent consultant who specializes in solving power problems. The judgment and opinions of a qualified, independent consultant are normally acceptable to both the end user and the equipment manufacturer/supplier.

Care should be used in selection of the consultant to ensure that he or she has experience in solving power problems for sensitive electronic equipment and does not have a vested interest in the recommended solution. For example, vendors of power conditioning equipment may have significant experience in solving power problems but their recommended solutions may be biased toward their product line.

6.3.4 The Electrical Contractor or Facility Electrician. Although an electrician may have a good working knowledge of power distribution systems, it will be generally ineffective to use a contractor for site surveys and analysis. Electricians normally do not possess the specialized equipment or expertise needed to clearly define the problem areas. However, an electrician can be a valuable assistant in the verification of the power distribution and grounding system for the layout of the power system feeders, branch circuits, and panelboards. In addition, would also be responsible for retrofit work to improve the facility wiring and grounding system.

6.3.5 The Electric Utility Company. An effective site power survey should include the involvement of the local electric utility. Utility personnel can provide site-specific information on disturbances (e.g., capacitor bank switching, reliability) that can occur on the utility system. Many power company engineers have expertise on effects of power quality problems on sensitive electronic equipment.

It is important to involve someone familiar with the local power system and the various factors that affect power quality from location to location. The utility engineer can fill this role in evaluating which disturbances may occur on the utility system and which protective equipment may be required by the user. Potential changes to the utility system that may improve power quality can also be evaluated. Some electric utilities offer preliminary site surveys prior to construction of facilities or installation of sensitive electronic loads. The monitoring equipment used can provide useful data on power disturbances at the point of interconnection. A growing number of utilities offer in-depth site surveys to pinpoint the source of power disturbances and provide assistance in selection of the appropriate power conditioning equipment. In many areas, electric utility companies have recognized the importance of power quality and are taking an active role in helping their customers solve power-related problems.

6.4 Conducting a Site Survey. Site surveys and analysis can be conducted in various levels of detail depending on the magnitude of the problem, amount of data desired, and economic factors.

A recommended breakdown of site survey levels is as follows:
(1) *Level 1 survey*. Testing and analysis of ac distribution and grounding system supplying the equipment.
(2) *Level 2 survey*. Level 1 plus monitoring of ac voltage supplying the equipment.
(3) *Level 3 survey*. Levels 1 and 2 plus monitoring of site environmental parameters.

When the desired level of the survey has been determined, the proposed analysis of results must be defined before any testing or power monitoring is initiated. Specific types of instruments are designed to detect specific problems and no single instrument has the capability to detect all types of problems. For example, a power monitor is designed to detect problems in the quality of the ac voltage; it will not detect wiring nor grounding problems. Unless the quality of the wiring and grounding system is tested and verified, the data produced by a power monitor are practically useless.

To conduct a site survey effectively, problem areas should be subdivided into at least three categories:
  (1) The condition of the ac distribution and grounding system.
  (2) The ac voltage levels of the power system.
  (3) The equipment environment, including temperature, humidity, electrostatic discharge, and radiated electromagnetic interference (EMI).

The order in which these categories are analyzed is critically important. Power distribution and grounding should be tested and analyzed before any testing is conducted to determine quality of the ac voltage.

6.4.1 Condition of the Power Distribution and Grounding System. Problems in industrial/commercial premises wiring and grounding account for a large share of all reported power quality problems. The greatest number of wiring and grounding problems is in the feeders and branch circuits serving the critical loads. The first activity in checking for power problems is to survey the soundness of the ac distribution and grounding system supplying the equipment. Problems in this category include such items as missing, improper, or poor quality connections in the power wiring and grounding from the source of power to the load. They can be generally classified as mechanical problems. Through error or oversight, intentional or unintentional, the power distribution and grounding system in many cases is not installed in accordance with the requirements of national, state, or local electrical codes and other specifications. For example, the National Electrical Code (NEC) (ANSI/NFPA 70-1993 [1][18] only permits a neutral-ground bond at the source of power (main panel or isolation transformer secondary), yet improper neutral-ground connections are a common problem encountered on power systems in the field. Experience has shown that many electronic equipment installations experiencing malfunctions and failures have one or more problems in the wiring and grounding system supplying the equipment.

In new installations, connections may be left off or not properly tightened. Reversal of conductors can also occur. Once the installation has been placed in service, vibration can loosen connections. Loads cycling on and off create heating and cooling that can eventually result in poor quality (high-impedance) connections. Also, periodic additions or modifications to the distribution system can result in missing, improper, or poor quality connections.

---

[18] The numbers in brackets correspond to those of the references in 6.8; when preceded by the letter "B," they correspond to those in the bibliography in 6.9.

Branch circuits are of lower power rating and are open to a greater variety of construction techniques and retrofit options, many of which cause problems. Caution should be exercised in the selection of test instruments used to conduct a verification of the power and grounding system. Use of the commonly available three-light circuit tester is not recommended and should be discouraged. These devices have some severe limitations and can provide a "correct" indication when the circuit being tested actually has one or more problems. In addition, they are incapable of indicating the integrity of the power conductors.

See Chapter 5, Table 5-1 for a discussion of recommended instruments to conduct the site survey.

6.4.1.1 Safety Considerations. Safety considerations come first in making any measurement on energized power systems. Instruments should be grounded using the manufacturer's recommendations. Continuity measurements should be made on de-energized circuits. Some measurements may require the use of a licensed electrician.

CAUTION! **Workers involved in opening energized power panels are required to abide by the prescriptions of ANSI/NFPA 70E-1988 [B4] concerning appropriate protective equipment, as well as government regulations codified in OSHA CFR 1910 [B5] and 1926 [B6], and in ANSI C2-1990 [B3].**

6.4.1.1.1 Neutral-Ground Bond. The neutral and equipment grounding conductor are required by the NEC [1] to be bonded at the main service panel and at the secondary side of separately derived systems. Improper, extraneous neutral-ground bonds are a relatively common problem that not only create shock hazards for operating personnel but can also degrade the performance of sensitive electronic equipment. These bonds can be detected using a wiring and grounding tester designed for the purpose. A voltage measurement between neutral and ground at the outlets will normally indicate voltage in the millivolt range under normal operating conditions. A reading of zero volts can indicate the presence of a nearby neutral-ground bond. Excessive current on equipment grounds in distribution panels also indicates the possibility of a load side neutral-ground bond. Visual inspection is necessary to verify and locate the bonds.

6.4.1.1.2 Measurements for Neutral Conductor Sizing. Measurements of load phase and neutral currents should be made to determine whether the load is sharing an neutral conductor with other loads and whether the neutral conductor sizing is adequate. For three-phase circuits supplying single-phase loads that have nonlinear current characteristics and share a common neutral, current in the neutral can exceed current in the phase conductor. This should be taken into account when sizing neutral conductors. Phase and neutral conductor measurements must be made with a true rms clamp-on ammeter to avoid inaccurate readings.

To determine whether the neutral serving the sensitive electronic load is shared with other loads, measure the neutral current with the sensitive load turned off. If the current is not zero amperes, a shared neutral is being used.

6.4.1.1.3 Measurements for Transformer Sizing. Transformers need to be sized according to load type. For nonlinear loads, or situations in which the load type cannot be determined in advance, Chapter 9 discusses transformer derating.

6.4.1.1.4 Equipment Grounding Conductor Impedance. Electronic equipment is required by the NEC [1] and local codes to be grounded through the equipment grounding conductor and bonded to the grounding electrode system at the power source. Impedance of the equipment grounding conductor from the electronic equipment back to the source neutral-ground bonding point is a measure of the quality of the fault return path.

Measure the impedance of the equipment grounding conductor using a ground impedance tester. An "open ground" indication reveals no equipment grounding conductor connection. A high-impedance measurement indicates poor quality connections in the equipment grounding system or an improperly installed equipment grounding conductor. Properly installed and maintained equipment grounding conductors will exhibit very low-impedance levels. Recommended practice is to verify an impedance level of 0.25 Ω or less. This will also help assure personnel protection under fault conditions [B9]. In many cases, with larger size equipment grounding conductors, the impedance should be much less than 0.25 Ω.

6.4.1.1.5 Neutral Conductor Impedance. Impedance of the neutral conductor from the sensitive electronic equipment back to the source neutral-ground bonding point is another important measurement. A low-impedance neutral is essential to minimize neutral-ground potentials at the load and reduce common-mode noise. The high levels of neutral current caused by nonlinear loads contribute to these problems.

The instrument used to conduct the equipment ground impedance measurements in 6.4.1.1.4 can also be used to measure the neutral conductor impedance. Neutral conductors should measure impedance levels of less than 0.25 Ω. High impedance in the neutral conductor can be the result of poor quality connections.

6.4.1.1.6 Grounding Electrode Resistance. The grounding electrode system is typically buried or inaccessible except during construction of the facility or major remodeling. The purpose of the grounding electrode system is to provide an earth reference point for the facility and provide a path for lightning and static electricity.

The resistance of the grounding electrode system should be checked at the time of construction. As a practical matter and for safety reasons, it is usually not measured again. In order to make the measurement accurately, the grounding electrode system must be disconnected from all other earth grounds. For new construction, measure the resistance of the grounding electrode system with an earth ground tester using the fall-of-potential method.

The integrity of the grounding electrode conductor is important because it serves as the connection between the building grounding system and the grounding electrode system. Use a clamp-on ammeter to measure current flow in the grounding electrode conductor. Ordinarily there will be a small but finite current flow. A lack of current flow may be indication of an open connection. Current flow on the order of the phase currents indicates serious problems and/or possible fault conditions.

6.4.1.1.7 Continuity of Conduit/Enclosure Grounds.
Electronic loads are recommended to be grounded with a separate equipment grounding conductor. The termination of the equipment grounding conductor can be either in an isolated ground system, insulated from the conduit ground, or it can be terminated in the conduit ground system. Either termination is ultimately connected to the building ground system. Both the isolated ground and the conduit ground must terminate at the first upstream neutral-ground bonding point. Ground impedance testers can be used to measure the quality of both the isolated ground and conduit ground systems from the equipment to the power source. Routing of phase, neutral, and equipment grounding conductors through continuously grounded metallic conduit is recommended practice for electronic equipment performance in addition to meeting safety codes. Continuously grounded metal conduit acts like a shield for radiated interference.

6.4.1.2 Performance Considerations.
Recommended methods for the determination of performance related parameters are discussed below.

6.4.1.2.1 Dedicated Feeders and Direct Path Routing.
Critical electronic loads should be served by dedicated feeders with conductor routing in as short and direct a path as possible. In conducting a site survey, it is necessary to determine if this is the case. One means to verify a dedicated feeder is to measure phase currents with the critical loads turned off. Any current flow indicates the feeder is being used to serve other loads.

6.4.1.2.2 Separately Derived Systems.
Separately derived systems have no direct electrical connection between the output conductors and the input conductors. Separately derived systems are required by the NEC [1]) to have a load-side neutral-ground bond that is connected to the grounding electrode system. All equipment grounding conductors, any isolated grounding conductors, neutral conductors, and the metal enclosure of the separately derived system, are required to be bonded together and bonded to the grounding electrode conductor. Visual inspections and measurements with a ground impedance tester can be used to determine the quality of these connections.

6.4.1.3 Wiring and Grounding Verification Procedures.
The services of qualified personnel, when conducting verification and testing of the power distribution system, should always be utilized. Their services will be needed

to provide access to power panels and assist in conducting the tests with maximum safety. In addition, they may be able to provide valuable information (e.g., history, modifications) about the distribution system.

While conducting the testing program, close visual inspections of power panels, transformers, and all other accessible system components should be made. Loose connections, abnormal operating temperatures, and other such items that can provide clues to the quality of the distribution system are particularly important to note. A good point at which to start the distribution and ground testing is the main building service panel or supply transformer. If the quality of the earth ground system is questionable, an earth ground tester can be used to measure the resistance of this connection. Additional tests at this location should include measurement of rms voltage levels (phase-to-phase, phase-to-neutral, and phase-to-ground), current levels (phase, neutral, and ground), and verification of proper neutral-ground bonding.

From this point, each panel in the distribution system serving the equipment should be tested and verified. Tests should include voltages, currents, phase rotation, ground impedance, and neutral impedance. Verification should include proper isolation of the neutral conductor, proper conductor sizing, tightness of connections, and types of loads being served.

Upon completion of the panel testing and verification, all branch circuits supplying the sensitive equipment should be verified. These tests should include voltages, proper conductor termination (wiring errors), and the absence of neutral-ground and isolated ground shorts, as well as measurement of ground and neutral impedance levels.

The recommended practice is to develop a systematic method of recording all observations and test results. This will enable efficient data analysis as well as ensure that no tests are overlooked. Figure 6-1 illustrates a sample set of forms for recording test results.

SITE SURVEYS AND SITE POWER ANALYSES

Power Distribution Verification Test Data

System Type: _____

Site: _____

Date: _____

Location: _____

Contact: _____

Phone: _____/_____

Source Transformer:

kVA: _____   Primary Voltage: _____   Secondary Voltage: _____

Taps: #1___ ; #2___ ; #3___ ; #4___ ; #5___ ; #6___ ; #7___ ; #8___

    Tap Position: _____

Measured Voltages and Currents:

    Primary Voltage   Primary Current

        A-B_____   A_____

        B-C_____   B_____

        C-A_____   C_____

    Phase Rotation:_____   G_____

    Secondary Voltage   Secondary Current

  A-B_____   A-N_____   A_____

  B-C_____   B-N_____   B_____

  C-A_____   C-N_____   C_____

  Phase Rotation:_____   N_____

  N-G Bonded? Yes_____ No_____   G_____

Remarks _____

**Fig 6-1
Sample Set of Forms**

Data Summary: Power Distribution and Grounding

Location:_____ Date:___/___/___

Panel:_____Room:_____User:_____

Power Source:_____

Panel Description:

    Manufacturer_____   Model_____

    Total Poles_____   Amperes_____

    Main Disc: Y___ N___   Amperes_____

Total Branch Circuits: 1 Pole_____ 2 Pole_____ 3 Pole_____

Feeder Description:

Phase Conductors: Size_____ Color_____ Copper: Y___ N___

Neutral Conductor: Size_____ Color_____ Copper: Y___ N___

Ground Conductor: Size_____ Color_____ Copper: Y___ N___

Neutral Bus:
    Isolated Neutral Bus Installed?   Y   N
    Total Number of Neutral Conductors?   _____

Ground Bus:
    Isolated Ground Bus Installed?   Y   N
    Insulated Main Grounding Conductor?   Y   N
    Conduit Main Grounding Conductor?   Y   N
    Secondary Grounding Conductor?   Y   N
    Total Number of Ground Conductors?   _____

Panel Status:
    Minimum NEC Working Clearance?   Y   N
    Branch Circuits Correctly Labeled?   Y   N
    Panel Name and Feeder Displayed?   Y   N
    Panel Hardware Working Correctly?   Y   N
    All Wiring Freely Accessible?   Y   N
    Abandoned Wiring in Panel?   Y   N
    All Connections Checked and Tight?   Y   N

Remarks:_____
_____

Fig 6-1 *(continued)*
Sample Set of Forms

SITE SURVEYS AND SITE POWER ANALYSES

IEEE
Std 1100-1992

Data Summary: Power Distribution and Grounding

Location:_____Date:___/___/___
Panel:_____ Room:_____

Voltage Readings: A to B _____ A to N _____
                   B to C _____ B to N _____
                   C to A _____ C to N _____
                   N to G _____ N to IG _____

Current Readings: Ph. A _____ Neutral   _____
                   Ph. B _____ Isol Gnd   _____
                   Ph. C _____ Ground   _____

                    Phase Rotation:
Ground Impedance:_____   Neutral Impedance:_____

Remarks:_____
_____
_____

Fig 6-1 *(continued)*
Sample Set of Forms

Data Summary: Power Distribution and Grounding						
Location:_____Date:___/___/___					
Panel:_____ Room:_____					
Branch Circuit Loads					
Pos	CB Size	Load	Pos	CB Size	Load
1			2		
3			4		
5			6		
7			8		
9			10		
11			12		
13			14		
15			16		
17			18		
19			20		
21			22		
23			24		
25			26		
27			28		
29			30		
31			32		
33			34		
35			36		
37			38		
39			40		
41			42		
Remarks:_____					

Fig 6-1 *(continued)*
Sample Set of Forms

SITE SURVEYS AND SITE POWER ANALYSES

Data Summary: Power Distribution and Grounding

Location:_____ Date:___/___/___

Panel:_____ Room:_____

Branch Circuit Tests

Circuit #	Voltage	Wiring	N/G Short	Ground Z	Neutral Z

Fig 6-1 *(continued)*
Sample Set of Forms

6.4.2 Quality of AC Voltage. Upon completion of the power distribution and grounding verification portion of the site analysis, the next step is to determine the quality of the power system, ac voltage waveforms. Various studies ([B1], [B2], [B7], [B8], and [B10]) have been conducted to quantify the types and frequency of occurrence of power line disturbances on circuits supplying sensitive electronic equipment. Generally, voltage disturbances as recorded by power line monitors can be classified into the basic groups shown in Chapter 3.

6.4.2.1 Detection of Voltage Disturbances. This section discusses the methods of detection for the various types of voltage disturbances included in Chapter 3. Recommendations for correction of these disturbances are covered in Chapter 9. A recommended practice is to periodically connect the power monitor to a disturbance generator and create known disturbances. Becoming knowledgeable on how the monitor reports these events will be helpful in the interpretation of data tapes from actual field sites.

6.4.2.2 Power Monitor Connections. Hookup of the monitor is an important consideration. If multiple channels are available, they should all be used to maximize the data obtained, enabling improved analysis of the number and the types of disturbances that have occurred. This analysis can then be applied toward the correct selection of power conditiong equipment to eliminate the problems. Figs 6-2, 6-3, and 6-4 illustrate suggested hookups for various power systems.

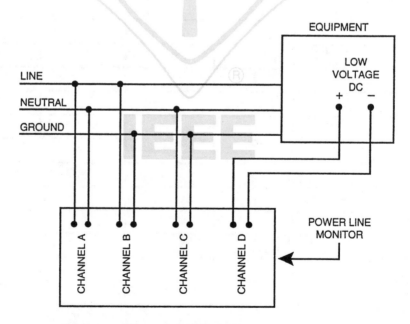

Fig 6-2
Recommended Power Monitor Hookup Procedure for
Single-Phase Applications

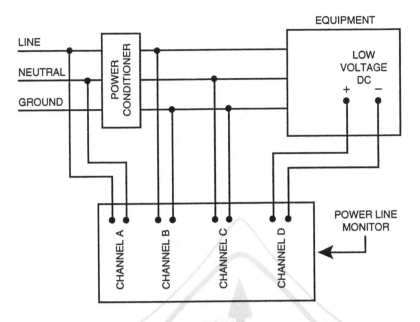

**Fig 6-3
Recommended Power Monitor Hookup for
Single-Phase With Power Conditioner**

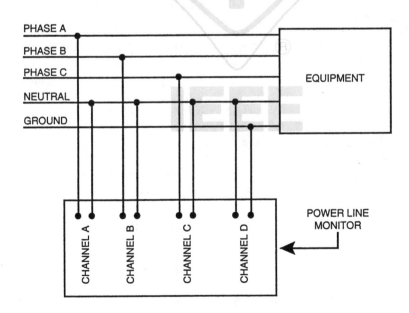

**Fig 6-4
Recommended Power Monitor Hookup for Three-Phase Wye**

A technique that can be used to determine what, if any, effect disturbances have on equipment is to connect the dc channel of the monitor directly to the output of the equipment power supply. Events occurring on the dc channel can then be correlated to events occurring on the input ac channels in determining the level of the disturbance at the sensitive logic circuits. When connecting the monitor to a power panel, always use a licensed electrician to make the connections. Assure that the connections will remain secure for the duration of the monitoring period.

NOTES: (1) Using twisted pair cables for monitor inputs will reduce the possibility of picking up radiated RFI/EMI fields.
(2) Connect monitor in the same mode in which the equipment is connected (phase-to-phase or phase-to-neutral).

6.4.2.2.1 Monitor Input Power. Recommended practice is to provide input power to the monitor from a circuit other than the circuit to be monitored. Some manufacturers might include input filters or transient voltage surge suppressors on their power supplies that can alter disturbance data if the monitor is powered from the same circuit that is being monitored.

6.4.2.2.2 Monitor Grounding. Care must be used in grounding of the monitor. Since a chassis ground is provided through the ac input power cord, any chassis ground connections to the circuit being monitored can create ground loops that result in additional noise being injected on the sensitive equipment feeder. To avoid this problem, it is recommended that no chassis ground connection be made to the circuit being monitored. The instrument manufacturer should be contacted for guidance as required.

6.4.2.3 Setting Monitor Thresholds. Once the hookup of the monitor has been determined, the next step is the selection of thresholds at which disturbances will be recorded. High and low thresholds should be set slightly within the voltage operating limits of the equipment. This will permit detection of voltage levels close to the critical maximum or minimum voltage limits that can result in equipment overstress or failures. If equipment tolerance limits are unknown, a high threshold of 125 V, and a low threshold of 105 V, is recommended for monitoring 120 V circuits.

Transient thresholds should be set to detect transients that cause component degradation or destruction. If no equipment transient limits are specified, a threshold of approximately 100 V-peak could be used. If the monitor has high-frequency noise detection, a threshold of 2–3 V-peak should be used for detection of high-frequency noise between neutral and ground.

Information such as the site, name, date, circuit being monitored, hookup scheme, and other related data, should be recorded at the beginning of the data printout to facilitate future reference to the data. Most monitors have the ability to be accessed via an RS-232 port by a remote terminal or computer. This feature can be very helpful in outputting data, changing thresholds, and performing other functions on several monitors in the field from a single terminal in the office.

6.4.2.4 Monitor Location and Duration. When monitoring a site that is serving several loads, it may be advantageous to initially install the monitor at the power panel feeding the system to obtain an overall profile of the voltage. The monitor can then be relocated to the circuits serving individual loads, such as central processing units (CPUs), disk drives, or other such loads that are experiencing malfunctions and failures. Comparison of disturbance data can provide clues as to the source of the disturbances and how to most effectively remedy the problem. It is generally recommended that the minimum monitoring period include at least one full work cycle, which would normally be seven or eight days. Longer monitoring periods are often needed to record disturbances that occur on a random or seasonal basis.

6.4.2.5 Analysis of Recorded Voltage Disturbances. Perhaps the most difficult task in conducting a site power survey is the analysis of the data provided by the power monitor. These data will be used in determining the source of the disturbances as well as making decisions on cost-effective methods for correction or elimination of the disturbances.

The individual responsible for the interpretation of data must have a thorough understanding of the disturbance capture and reporting characteristics of the specific monitor used in the site survey to minimize the possibility of misinterpretation. One of the factors to be determined is whether a particular disturbance is causing an equipment malfunction. This relationship is relatively easy to determine if an equipment malfunction occurred at the same time the disturbance was recorded.

In many cases, disturbances are recorded and appear to have no effect on equipment performance. These disturbances could still be severe enough to cause degradation of components that eventually results in premature failure. Part of the data analysis is a determination of the source of the disturbances, which can prove to be a very elusive task. Disturbances can be caused by the equipment itself, by other equipment within the facility, by equipment external to the facility, by power utility operations, by lightning, or any combination of these sources. Although a complete description is not possible in this Recommended Practice, some general guidelines can be helpful.

If the equipment is supplied by an isolation transformer or a power conditioner and disturbances are recorded on the output of the conditioner only, then the conditioner or the equipment itself may be the source.

Compare disturbances on the dc output of the power supply to events on the ac input to the equipment. If no time correlation can be made, the events on the dc channel could be originating at an external device and being reflected into the system by the data or communication cables. If disturbances are occurring about the same time during the working day, try to determine what equipment is being operated in the facility at those times. If no correlation can be obtained, then the source may be external to the facility.

Disturbances that occur at exactly the same time each day are caused by equipment that is time-clock-controlled. One such type of equipment is switched capacitor banks used by power utilities. Contacting the power utility company to determine what operations are being conducted on their system at various times of the day can often provide helpful information.

6.4.3 Electronic Equipment Environment. Electronic equipment malfunctions and failures can be caused by improper environmental parameters such as temperature, humidity, EMI, and electrostatic discharge (ESD). A site survey should include testing or monitoring of these parameters to confirm a proper environment for the equipment.

6.4.3.1 Temperature/Humidity. Most monitors that are used to measure voltage disturbances have transducers available to measure temperature and humidity. Once the temperature and humidity specifications from the equipment manufacturer have been obtained, set the high- and low-threshold points slightly within those limits in order to capture variations that are close to the limits of the electronic equipment. Recommended practice is to program the monitors so that long-term (12- or 24-hour) reports of temperature and humidity levels are documented. Compare any sudden changes in temperature and humidity to the site error-logs to see if any correlation can be made. High levels of temperature can cause overheating and premature failure of components. High humidity can cause condensation resulting in intermittent contacts on circuit boards. Low humidity can be a contributing factor causing increased levels of electrostatic discharge.

6.4.3.2 Electromagnetic Interference (EMI). Radiated EMI can impact the performance of sensitive electronic equipment. In attempting to confirm whether the problem is EMI, the first step is to establish the method of site operations. Are any transmitters or other communication devices being operated near the electronic equipment? Can correlation be made between the radio operation and equipment malfunctions? A visual inspection of the surrounding area can be conducted looking for external sources of EMI such as radio/TV towers, microwave towers, airports, etc. Generally, two levels of EMI measurements can be conducted. The first is measurement of high-frequency fields using a field-strength meter or EMI transducer coupled to a power monitor. This technique is recommended as a preliminary step to either confirm or eliminate EMI as a problem. Consult the electronic equipment manufacturer for the equipment susceptibility limits. If excessive levels of radiated fields are indicated, recommended practice is to conduct a complete EMI survey using a spectrum analyzer, which is the second level of EMI measurement. This survey is intended to pinpoint the frequency and direction of the signal source so that corrective measures can be taken. Recommended corrective measures for EMI problems include the following:
 (1) Reorienting or relocating the sensitive equipment or source,
 (2) Removal of the source, and
 (3) Shielding of the source or affected equipment.

6.4.3.3 Electrostatic Discharge (ESD). ESD can severely impact the performance and reliability of electronic equipment. A site can experience failures from ESD and not immediately be aware of the problem since voltage levels that can cause component failure are below the perception threshold of the individual. Meters are available to measure the level of static charge on personnel and equipment. Recommended practice is to measure static charge on personnel, furniture, and other such items located in the vicinity where the sensitive equipment is being operated. If equipment failures are caused by ESD, recommended corrective measures include the following:

(1) Maintaining proper humidity levels in the equipment areas,
(2) Using antistatic mats on floor and work surfaces,
(3) Replacing static generating items, such as chairs, that aggravate the ESD problem, and
(4) Training operating personnel to discharge themselves before operating the sensitive equipment.

6.5 Applying Data to Select Cost-Effective Solutions. Upon completion of the field testing and power monitoring portion of the site survey, it is recommended that all data be classified into distinct categories before analysis. This will assist in defining problem sources as well as identifying means of correction.

For example, a high-impedance neutral conductor on the incoming feeder to a power panel may be the cause of common-mode noise that is being reflected into the entire system. Since distribution and grounding problems are mechanical (loose), missing, or improper connections, the means of effective correction is also mechanical (a screwdriver). Thousands of dollars have been spent on conditioning equipment in an attempt to solve a power problem that was finally corrected with a screwdriver. It is recommended that problems found in the power distribution and grounding system be corrected before attempting correction of problems in the quality of ac voltage. These distribution problems can normally be remedied at minimal cost, and in some cases may be the only correction needed to assure a high degree of system performance and reliability.

Careful analysis of the power monitoring data is necessary to determine the types, quantity, and severity of the disturbances recorded as well as the immediate or long-term impact on equipment performance and reliability. It is these data that will form the basis for making decisions as to what type of power conditioning equipment will be required to eliminate the problem. A discussion of the various types and applications of power conditioning equipment is provided in Chapter 8.

6.6 Long-term Power Monitoring. Studies have been conducted using power monitors to determine the quantities and types of disturbances that occur over an extended period of time ([B1], [B2], [B7], [B8], and [B10]). A comparison of three of these studies is included in Appendix A.

Although these studies can provide some helpful information, caution must be exercised in applying this information to correct problems on a specific site. Numerous variables enter into the equation that determine the types and

quantities of voltage problems occurring on any given site utilizing sensitive electronic equipment. They include the following:
(1) Type and configuration of the electronic system installed (e.g., data processing, telecommunications, process measurement and control, point-of-sale terminals).
(2) Configuration and condition of the power distribution and grounding system supplying the equipment.
(3) Quantity, location, and type of power protection equipment installed.
(4) Other equipment operating from the same power distribution system in the facility.
(5) Location of the facility on the utility power system.
(6) Other facilities in the immediate area served from the same power utility system.
(7) Geographic location of the facility (exposure to lightning).

6.7 Conclusions. Conducting a site power analysis or site survey can be an effective means of detecting and correcting power-related problems if it is properly applied. Careful testing and troubleshooting techniques are necessary to collect power quality data that is meaningful. Classification and thorough analysis of all data must be conducted in order to define as accurately as possible the problem areas.

Understanding the advantages and limitations of each type of power conditioning equipment will enable the selection of the most cost-effective conditioner that will solve the problem. Also note that even power conditioning equipment can be defeated by improper grounding. Since the responsibility for correction of power-related problems lies with the customer or end-user, it is critically important the specific recommendations for correction are clearly communicated to avoid needless expenditure of funds on conditioning equipment that may not correct, or can even aggravate the problem.

6.8 References. This standard shall be used in conjunction with the following publication. When the following standard is superseded by an approved revision, the revision shall apply:

[1] ANSI/NFPA 70-1993, National Electrical Code.[19]

6.9 Bibliography

[B1] Allen, G. W. and D. Segall, "Impact of Utility Distribution Systems on Power Line Disturbances," *IEEE Summer Power Meeting Conference Paper*, SUMPWR-76, A76-338-4, 1976 (abstract in *IEEE Transactions on PAS*, Vol. PAS-95, Nov/Dec 1976, pp. 1760–61).

---

[19] NFPA publications are available from Publications Sales, National Fire Protection Association, 1 Batterymarch Park, P.O. Box 9101, Quincy, MA 02269-9101, USA.

[B2] Allen, G. W. and D. Segall, "Monitoring of Computer Installations for Power Line Disturbances," *IEEE Winter Power Meeting Conference Paper*, WINPWR C74 199-6, 1974 (abstract in *IEEE Transactions on PAS*, Vol. PAS-93, Jul/Aug 1974, p. 1023).

[B3] ANSI C2-1990, National Electrical Safety Code.

[B4] ANSI/NFPA 70E-1988, Standard for Electrical Safety Requirements for Employee Workplaces.

[B5] CFR (Code of Federal Regulations), Title 29, Part 1910: Occupational Safety and Health Standards (OSHA).

[B6] CFR (Code of Federal Regulations), Title 29, Part 1926: Safety and Health Regulations for Construction (OSHA).

[B7] Goldstein, M. and P. D. Speranza, "The Quality of U. S. Commercial ac Power," *INTELEC (IEEE International Telecommunications Energy Conference)*, 1982, pp. 28–33 [CH1818-4].

[B8] Key, T. S. "Diagnosing Power Quality Related Computer Problems," *IEEE Transactions on Industry Applications*, Vol. IA-15, No. 4, July/Aug 1979.

[B9] Kleronomos, Chris C. and Edward C. Cantwell, "A Practical Approach to Establish Effective Grounding for Personnel Protection," *IEEE Industrial and Commercial Power Systems Technical Conference (I &CPS)*, 1979, pp. 49–57 [CH1460-5].

[B10] Martzloff, F. D. and T. M. Gruzs, "Power Quality Surveys: Facts, Fictions, and Fallacies," *IEEE Transactions on Industry Applications*, Vol. 24, No. 6, Nov/Dec 1988, pp. 1005–18.

Chapter 7
Case Histories

7.1 General Discussion. This chapter presents case histories involving typical problems that have been encountered in the powering and grounding of sensitive electronic equipment. It is hoped that readers will find that the solutions presented will be helpful in solving problems being encountered currently.

Since most of the case histories in this chapter can fit into two or more subject areas, it is suggested that readers peruse related topics as presented in the subsections that follow.

7.2 Typical Utility-Sourced Power Quality Problems. Many utilities are located in areas where lightning storms commonly occur during portions of the year. These storms are notorious for causing ground faults on both transmission lines and ac distribution lines. These faults often cause nearby utility-fed electrical equipment to power-down.

Utilities must also maintain relatively constant distribution voltages, as their customers' load mix and total power demand vary enormously. In an effort to maintain constant voltage, utilities switch large capacitor banks into and out of their transmission and distribution systems. This switching may result in voltage surges at customer sites.

Case histories of these two primary sources of utility power disturbances are presented below. It should always be emphasized that personnel at sensitive electronic facilities must have a good working relationship with a knowledgeable power quality consultant and with their utility company. Involvement of utility company personnel should not be adversarial, for much can be accomplished by working together to solve the problems.

7.2.1 Voltage Sags Due to Utility Fault Clearing. Several real-life events can induce phase-voltage to ground (or neutral) shorts on the utility ac transmission and distribution system. Thunderstorms, high-wind conditions, and small animals are the primary causes of these short-duration events. The typical utility response is to first re-energize the affected circuit(s). This activity often is successful in clearing the initial line-fault, but also produces a momentary sag for other users on the same ac distribution system. The following cases studies exemplify this problem.

Problem 1: Nuisance Production Interruptions During Thunderstorms. A plastics film extruding facility was experiencing numerous production interruptions during thunderstorms. Detailed monitoring and analysis showed that adjustable frequency drive (AFD) motors automatically tripped off-line for voltage sags below 85% of nominal. Equipment downtime logs correlated with the local utility's logs of momentary short circuits on their distribution power line.

Solution. The AFDs were modified for a lower ac input voltage range (changed input threshold from 85% to 75% of nominal voltage). The plant ac distribution voltage, from the stepdown transformer supplying the affected equipment, was increased (via changing its tap setting) to a level that allowed the equipment to ride through many of the voltage sags. This voltage change was accomplished without exceeding the maximum allowable voltage (per ANSI C84.1-1989 [1][20]) under normal conditions.

Problem 2. Excessive Assembly-Line Restart Times After Momentary AC Sag Conditions. An automobile transmission plant experienced numerous partial stoppages in its production line during thunderstorms. Excessive time was required to restart the affected production equipment (solid-state controllers) and to balance product flow through the total production line. Detailed monitoring and analysis showed that the most sensitive programmable logic controllers ceased functioning when the ac line voltage dropped below 87% of nominal for periods longer then 8 ms. Utility analysis showed that the median voltage of utility originated sags was 65% of nominal, with a median duration of 190 ms.

Solution. Since the median sag voltage was within the operating range of several voltage-regulating transformers, the power quality consultant recommended that voltage-regulating power conditioners be added to power the less sensitive machine tool controllers, and that small, dedicated uninterruptible power supply (UPS) be used to power the critical controllers.

7.2.2 Voltage Surges Due to Utility Power-Factor/Voltage-Regulation Capacitor Switching. Utilities often find it necessary to add/subtract capacitance to their ac transmission and distribution system to achieve reasonable power factors and voltage levels. When these capacitor banks are switched to and from the ac distribution system, they create phase-voltage surges. These surges appear as decaying-oscillatory surges to the user. Examples of the impact of these surges follow.

Problem 1: Nuisance Production Shutdown of Steel Mill Casting Process. This case involves the AFD motors at a steel mill continuous casting plant. The AFD motors drive slabs of molten steel through the casting process. When the AFDs tripped out, molten steel solidified within the casting machines and resulted in considerable production downtime. Detailed monitoring and analysis showed that ac line surges on the feeder to the AFDs had a 600–700 Hz ringing frequency, 0.6 pu (per unit) initial amplitude, and durations of 8–10 ms. The 6900 V feed to the building showed identical surges.

Solution. The utility added a preinsertion-inductor to their capacitor bank. This inductor is momentarily placed in series with the capacitor bank, when the capacitor is switched into the ac distribution system. The inductor prevented further AFD trip-outs, due to utility capacitor switching. The

---

[20] The numbers in brackets correspond to those of the references in 7.11.

switching surge at the AFD (with the preinsertion inductor) had an initial amplitude of 0.2 pu and 1600 Hz ringing frequency.

Problem 2: Intermittent Shutdowns of Adjustable Frequency Drives (AFDs) in a Manufacturing Plant. This case involves a manufacturing plant, located in the southern United States, where 5 hp AFD motors frequently would trip-out intermittently. The result of these disruptions, on a multistage continuous-processing line, was considerable material loss and excessive line restart time. The AFD diagnostic code typically would indicate an "overvoltage on the ac power" feeding the system. Steady-state, true rms, voltage readings on the 480 Vac, three-phase line were on the low side at 452–479 Vac. Further measurements correlated voltage surges with utility-level power-factor correction capacitor switching.

Solution. Consultation with the AFD manufacturer revealed that the AFD protection circuitry was sensitive to ac overvoltage conditions for extremely short time periods (e.g., 800 V for 40 µs). The manufacturer solved the problem with the addition of transient voltage surge suppressors, which clamped the ac voltage to less than 750 Vac.[21]

Problem 3: Repeated Computer Lockup and Component Failures in CT Scan Equipment at a Medical Clinic. Monitoring input of the 480–208Y/120 Vac, three-phase transformer, feeding the CT scan equipment, revealed surges characteristic of utility power-factor correction switching (decaying-oscillatory waveform). Monitoring of the output of the same transformer showed a reduced surge, but still sufficiently severe to cause operational failure of the CT scan equipment.

Solution. Detailed analysis of the surges showed a characteristic ringing frequency of 1.0–1.5 kHz. This allowed the specification of an active-tracking filter, specifically designed for this type of surge, to be installed.

7.3 Premises Switching Generated Surges. The majority of ac voltage surges experienced at intermediate-sized and large user sites have been found to originate within the site itself. Switching of reactive loads is the primary cause of such surges. Below are examples of these switching surges.

Problem 1: General Case of Voltage Regulation Capacitors on the Secondary of Service Transformer. Voltage regulation of large industrial facilities can be accomplished through the application of one or more capacitor banks on the secondary side of the service transformers. The capacitors are voltage controlled; i.e., they are energized when the (utility) delivery voltage drops to a certain value, and they are de-energized when the voltage rises to a predetermined setting.

A typical design is to size these capacitor banks to provide a voltage boost of approximately 2%. The energization of these capacitor banks cause a decaying oscillatory surge whose peak value can reach 1.8 pu (per unit). The frequency of these surges can be closely approximated from the percent boost:

$f_{surge} = 60 / [\text{boost (pu)}]^{1/2}$

---

[21] The classical solution to this problem, which is more broadly applicable, is to provide a series reactance on the ac supply to the AFD system.

Thus, a capacitor sized to provide a 2% boost will have an oscillatory frequency of approximately 420 Hz.

Since no three-pole switching device is restrike-free, capacitor switching surges can reach 2.5 pu or more, following one or more restrikes.

In the past, these transients did not seem to cause problems to plant equipment, which were predominately lights and motors. Because of the rapid evolution of semiconductors, practically every mechanical device requiring automatic control now uses some type of solid-state control system. More and more industrial plants are relying totally on solid-state and microprocessor-based equipment for data processing, communications, and process controls.

Solution. Transient voltage surge suppression (TVSS) devices are installed to protect sensitive electrical equipment. The protection scheme includes surge protective devices, installed at the plant service-entrance as well as downstream at the sensitive electronic equipment.

Problem 2: Notching Surges on the AC Distribution System. After installing a 1000 hp solid-state dc drive, a manufacturing plant began to have operating problems with existing solid-state equipment, connected to the same 480 V, three-phase distribution system. The distribution system supplying the dc drive was an ungrounded system. Operational problems of the nearby electronic equipment were attributed to the line-voltage notching surges. These surges were caused by the new solid-state dc drive.

Line-voltage notching surges are produced during motor-control commutation. When the current flowing in one phase must suddenly stop and this same current must suddenly start flowing in a different phase, a voltage surge occurs. Since the line has finite inductance, which prevents instantaneous changes in current flow, a momentary short-circuit occurs between phase and ground during this transition.

Solution. A shielded isolation transformer was installed on large dc drives to isolate their commutation-produced ac line voltage notching from other voltage sensitive solid-state equipment. The shielded isolation transformer acted to reduce the commutation-induced surges, and resultant ground current flows, to levels acceptable for the surrounding electronic loads.

Problem 3: Solid-state Adjustable Frequency Drives (AFDs). A manufacturing facility could not keep their AFD motors running. The input fuses on the AFDs were open-circuited on a regular reoccurring basis. A log of the AFD fuse failures was made. This was compared to the utility outages and other external causes. The result of the comparison indicated numerous unaccounted fuse failures. A digital power disturbance analyzer was installed and used to indicate the power quality at the drives.

The disturbance analyzer indicated a high concentration of power disturbances. These disturbances were mostly phase-neutral voltage surges, originating from within the building. Significant neutral-ground voltages were also observed. The disturbances were being generated primarily by the AFDs themselves, and causing other nearby AFDs to malfunction.

Solution. Shielded isolation transformers and TVSS devices on each AFD solved the problem.

CASE HISTORIES

7.4 Electronic Loads. Electronic loads have in common the characteristic of generating harmonic currents. These harmonic currents circulate within the ac distribution system, which supplies power to the electronic loads. These harmonic currents contribute to I^2R heating within the ac distribution system and can cause considerable voltage waveform distortion. Below are case studies that deal with harmonic currents and their impacts on the ac distribution system.

7.4.1 Uninterruptible Power Supply—Unfiltered Input. Static uninterruptible power supplies are nonlinear loads on the ac distribution system, and as such they generate and feed back harmonic currents into the ac distribution system. These harmonic currents can negatively impact other sensitive electronic equipment on the same ac distribution system.

Problem 1: Office Copier Lock-up and Laser Printer Data Errors. A southwestern university constructed a new classroom facility near two existing (large) computer room UPSs, and fed the new facility from the same ac distribution that fed the UPSs. Office equipment, with high-current-consuming electronically controlled heaters in the new classroom facility, experienced repeated data errors and often lock-up. User personnel were able to make the office equipment temporarily operational by switching off and on the ac power to the equipment. No equipment damage occurred.

After minor miswirings were corrected in both the neutral and equipment grounding conductors, power quality analyzer results showed that severe phase voltage notches (from the UPS) and common-mode noise voltages ranged from 5–35 V. Initial attempts to solve the problem with a nonoptimized (50 dB) filtering circuit, normal- and common-mode TVSS, a shielded isolation transformer, and a dedicated (line interactive) UPS failed.

Solution. Additional waveforms captured by the power quality analyzer showed that the line interactive UPS did succeed in eliminating the phase voltage notches, but since its design did not bond its output neutral with the equipment grounding conductor, it was ineffective in reducing the common-mode noise. Conversely, the shielded isolation transformer was shown to be effective in reducing the common-mode noise, but not the phase voltage notching. Armed with this data the power quality consultant recommended that a device be employed that could both regenerate the phase voltage and establish a local neutral-ground bond. Installation (near the sensitive equipment) of a simple ferroresonant transformer, with its secondary neutral bonded to the equipment grounding conductor, and TVSS, allowed for error-free operation of the office equipment.

7.4.2 Uninterruptible Power Supply—Unfiltered Output. Static UPSs, not equipped with sufficient output filtering, supply their loads with harmonic-rich voltage waveforms, and are generally unable to fully filter harmonic currents generated by their electronic loads. This situation may result in both the misoperation of load equipment and overheating of components in the UPS output distribution system.

Problem 1: Semiconductor Production Test Equipment Malfunction. A major semiconductor manufacturer was experiencing erratic yields from a group of wafer-level production testers. Initially, power issues were not suspected due to the testers being supplied from a UPS.

Power analyzer indications that the UPS was supplying power with high-voltage harmonic content, to the test equipment, were confirmed when the equipment fed was switched to an adjacent utility source and the tester immediately started logging repeatable results.

Solution. UPS manufacturer added a 5% THD harmonic filter to the existing installed UPS, and showed the same repeatable test results as attained when the tester was powered by normal utility power.

7.4.3 Automated Office. Automated offices have, as a significant fraction of their total electrical power consumption, nonlinear load equipment. This includes such equipment as computers, terminals, printers, copiers, and facsimile machines. All the triplen-harmonic currents generated by this equipment are returned on the neutral conductors (between the electronic loads and their separately derived source) of the ac distribution system. Historical practice has been to reduce the size of ac distribution neutral-conductors. Present usage patterns of automated office equipment can be at odds with the original ac distribution system design assumptions, which may result in misoperation of sensitive electronic equipment and excess heating of ac distribution elements.

Problem 1: Intermittent Data Memory Errors and Data Transmission Errors Between Remote Terminals and a Central Computer System. In the installation of a multistory office building, several floors of modular workstations were powered from a common, three-phase, step-down transformer with a shared neutral. Mainframe terminals in these workstations experienced intermittent data memory errors and data transmission errors, and occasionally experienced hardware failures.

The power quality analyzer showed that considerable (and variable) neutral current existed and that the neutral-ground voltage mirrored the neutral current flow. RMS values of neutral-ground voltage did not exceed 3.7 V, but the peak voltage ranged up to 10 V, which was well above the logic and signal voltage levels. The analyzer also showed phase voltage switching surges up to 800 V.

Solution. Each floor of the building was isolated into two sections via shielded isolation transformers, with TVSS. Individual neutral conductors were installed for all workstation branch circuits. Reestablishment of the neutral-ground bond at the new isolation transformers, combined with the reduced neutral current in the dedicated neutrals, reduced neutral-ground voltages to less than 2 V.

Problem 2: Engineering Laboratory with Offices Adjacent. Computer system used by laboratory personnel had intermittent failures and data errors. Typically, the misoperations would start around 10:00 A.M., but the timing was variable, and on some days there were no failures at all.

Power quality analyzer measurements showed very regularly timed repetitive sags in the phase voltage, beginning just after 10:00 A.M. Close analysis of the data showed corresponding surges between neutral-ground conductors. Additional investigation located a laser printer in a nearby office, whose "print fusing heater" switched on approximately every minute. The high current demand and resulting voltage developed between the neutral and equipment grounding conductors caused the computer system data errors.

Solution. The offending laser printer was moved to another branch circuit thereby eliminating its interaction with the computer system's neutral conductor.

7.4.4 Interaction between Power-Factor/Voltage-Regulation Capacitors and Electronic Loads. Harmonic currents generated within electronic loads can result in excessive heating of ac distribution and transmission components. In severe cases, the excess currents result in ac distribution or ac transmission system malfunctions.

Problem 1: Excessive Failures of Utility Capacitor Bank Over-Current Protectors. An industrial plant, supplied from two parallel transformers totaling 7500 kVA, has an 1800 kvar capacitor bank for voltage regulation. The electric utility had experienced two main problems with their capacitor bank:

(1) The capacitor circuit breaker overheated, with temperatures exceeding 70 °C, and
(2) Both 100 A individual capacitor can fuses and 3000 A main line fuses were blowing.

The harmonic content of the load current varied continuously. The 5th harmonic was by far the largest at 1080 A. This was approximately 23.5% of the 4600 A fundamental current.

Although the resonant frequency of the system is close to the 6th harmonic, the 5th harmonic current increased in the capacitor by a factor of 2.5, or 2700 A. With the 5th harmonic current alone, the total rms current in the capacitor was 64% higher than the fundamental current of 2077 A, and greater than the ampere rating of the capacitor fuses! These conditions were bad enough, but at the customer's peak load current of 6400 A (30% higher) the conditions were worse.

Solution. Since the plant personnel were unwilling to spend any money to reduce the 5th harmonic, the utility had to remove the capacitors. This resulted in the voltage regulation at the plant to be significantly worse. A more correct solution would have been to add filtering for the 5th harmonic current, and to assess the voltage capabilities of the existing power factor correction capacitors.

7.5 Premises-Wiring Related-Problems. Many of the power quality problems found in service industry facilities are associated with faulty premise wiring. Below are identified major types of premise-wiring-related problem areas.

7.5.1 Receptacle Level Miswiring

Problem 1: Miswired Receptacle on Personal Computer. An automobile parts store was having difficulty getting their personal computer to operate. The personal computer would frequently lock-up or experience other unexplained failures. A check of the utility company's outages did not reveal any correlation between them and the equipment problems. The power outlet serving the computer was examined for polarity, revealing that the phase and the neutral conductors were reversed at the receptacle.

Solution. Once the conductors were correctly connected, the personal computer in the automobile parts store ran without any problems.

7.5.2 Feeder and Branch Circuit Level Miswiring.

Problem 1: Ground Conductor on Outside of Conduit. A computer manufacturer required the electrical contractor to install #4 AWG insulated ground wire on the outside of the conduit, which carried its associated electrical power conductors. The insulated ground wire was tie-wrapped to the conduit. The manufacturer wanted the ground wire to be separated from the power conductors, yet he wanted to use the same ground wire. The electrical inspector rejected this installation on the basis that the impedance was increased. The National Electrical Code (NEC) (ANSI/NFPA 70-1993 [2]) requirement for a "sufficiently low impedance to . . . facilitate the operation of the circuit protective devices in the circuit" was not met.

Solution. The recommended practice is to have the ground conductor on the inside, not the outside, of the conduit. When this change was implemented, the electrical inspector approved the installation.

7.5.3 Ground-Fault Circuit Interrupter Problems

Problem 1: Capacitor Bank Switching. An industrial facility, served from a wye-connected 12.5 kV service and grounded through a resistor, had two 3600 kvar capacitor banks for voltage regulation. Plant personnel complained that a feeder circuit breaker would trip periodically from a ground fault relay. They would spend time inspecting the equipment downstream for ground fault but could never find the cause. After considerable investigation, the problem was discovered to be caused by the capacitor bank. Transient overvoltages created from capacitor switching would sometimes be high enough to cause a TVSS device on the high side of one of the downstream transformers to operate. The power-follow-through current through the TVSS device was enough to cause the ground relay to operate.

Solution. The easiest solution was to raise the current setting on the ground fault relay.

7.5.4 Ground Discontinuity

Problem 1: Lightning and Isolated Grounding Systems. In an area experiencing several short lightning storms a day, erratic computer operation was occurring within a large office building. The grounding conductors of all the computers in the building were insulated from all other conductors, bused together at a single location, and then carried out of the building to a grounding electrode system. This isolated grounding electrode system consisted of six rods driven into the earth away from the building. Power quality monitor measurements indicated that voltages were being developed between the building grounding system and the isolated computer ground. A code analysis showed that the grounding configuration was in violation of the NEC [2]), and was a definite safety hazard.

Solution. The computer manufacturer was persuaded to permit grounding of the computer system to the building grounding system, which had a concrete-encased grounding electrode. Immediately, the erratic operation of the computer system ceased, and the safety hazard was eliminated.

Problem 2: Multiple Buildings, Common Signal Conductors. Four separate but nearby buildings contained the control for a series of chemical processes. Each building housed a computer for process control. There was some interconnection between data circuits of the computers in adjoining buildings. Each of the four computers was separately grounded to the structural steel grounding system of its building. Operation of the computer systems was erratic.

Solution. The responsible plant engineer chose to replace the interconnecting datalines with fiber optic communication links. The optical fibers were buried in plastic conduits and the interface electronics in each building were grounded with their respective power grounding means.

Problem 3: Computer Graphics System—Ground Potential Difference on the Data Lines. A CAD/CAM graphics system, installed by a computer graphics vendor, links a central processing unit (CPU) to remote terminals in separate buildings. No direct lightning strikes were reported, yet extensive damage was done to the circuit boards on terminals and the CPU inputs. Power surges were suspected and precautions were taken, but they did not help. Isolation transformers were installed but, again, did not help.

The data cables had been run in plastic underground conduit with only one end of the shield grounded and the other end left floating. High voltages were produced between the floating end and its local ground. The problem was not in the power line surges, but the differential ground potentials. The opto-isolators did not isolate the problem since local power transformers to the opto-isolators closed the loop.

Solution. The solution was to tie the floating end to ground through a TVSS device. This allowed short rise-time surges to be conducted to ground, without establishing a ground loop. A better solution would be to use a dual-shielded cable, its outer-shield grounded at both ends and its inter-shield grounded at one end, with a TVSS device (connected to ground) at the other end. If data integrity is an absolute requirement, metallic connections should be avoided for data links spanning remote terminals.

Problem 4: Computer-Aided Industrial Control—Ground Potential Differential on Power Lines (Absence of Ground Window). A novel adaptive control system, using microprocessor-based sensors and phase-control of power thyristors, had suffered system crashes and memory component damage on repeated occasions. Suspicions developed that there were some correlations between the crashes or damage and the operation of another developmental power system in an adjacent laboratory.

A review of the total power system revealed the existence of ground loops. On one side, the power supply for the computer and some signal processing circuits were obtained from the room outlets of the laboratory 120 V system, including the grounding conductor (green equipment safety conductor). On the other side, the power supply for the high power circuit was obtained from a feeder coming directly from the building power center, including again a grounding conductor run alongside this power line. Everything was properly installed by electricians and bonded to the frame of the machine being controlled.

A double ground-loop was formed: one between the grounding conductor of the 120 V room supply and the power-feed grounding conductor, and the other between the signal processing ground and the voltage-probe with its separate ground reference.

During transient conditions involving the high power feed to this system and the neighboring system, substantial current could flow between the two ground wires linked by the computer signal wiring.

Solution. An immediate cure was to open the surge ground current path (between the two systems) by inserting a shielded isolation transformer in the 120 V supply to the computer and bonding the secondary side of this transformer to the single-point-ground derived from the high-power feed (an NEC [2] requirement). This correct application of an isolation transformer, to open a ground loop, is in contrast to the misconception that isolating transformers can eliminate line-to-line surges.

Clearly, the first ground loop was one of the major sources of the problem, which could have been avoided had the system been arranged with a ground-window.

7.6 Transient Voltage Surge Suppression Network Design

Problem 1. During the initial startup of a solid-state motor drive in a chemical processing plant, difficulties arose with the varistor and its protective fuse at the input of the thyristor circuits. Frequent blowing of the input power fuse was observed, with occasional failure of the varistor. The plant substation, fed at 23 kV from the local utility, included a large capacitor bank with one-third of the bank switched on and off to provide power factor and voltage regulation. These frequent switching operations were suspected of generating high-energy transients that might be the cause of the failure of the fuses and varistors, because literally thousands of similar drive systems have been installed in other locations without this difficulty. On-site measurements indicated that indeed the fuses and varistors were not matched to their environment.

Solution. Immediate relief was secured by the installation of a larger varistor at the same point of the circuit: long-term protection was obtained by the addition of a gapless metal-oxide varistor on the primary side of the step-down transformer feeding the drive.

This case history illustrates how surge protection devices that are successfully applied for the majority of cases can fail when exposed to exceptionally severe surge environments. It also shows how little attenuation occurs, at the frequencies produced by switching surges, between the distribution level, (23 kV) and the utilization level (460 V), even though a long line and two stepdown transformers exist between the source of the transient and the point of measurement.

In addition to the proposed upgrading of protection at the 460 V level, three other remedies could be considered: installation of surge arresters at the intermediate voltage level (4160 V), installation of surge arresters at the 23 kV level, or a change in the circuits involved in the capacitor switching, designed to reduce the severity of the transients at their origin.

(This solution can be considered site-specific, and the solution applies only to this problem at this site. Care must be exercised in applying this to other sites.)

7.7 Typical Radiated EMI Problems

Problem 1: Computer Monitor with Wavy Screen Distortion. A high-resolution computer monitor was experiencing a wavy/quivering screen distortion. Magnetic field measurements showed the 60 Hz field in the vicinity of the monitor was about on order of magnitude higher than elsewhere in the office. An inspection of the area revealed that a subpanel (magnetic field source) was located on the other side of the wall next to the monitor.

Solution. The user moved the monitor approximately 4 ft away from the back of the subpanel, where the magnetic field strength was at a nondisturbing level.

7.8 Typical Electrical Inspection Problems.

The incidents related are very brief. As far as the electrical inspector was concerned, these examples were open and shut cases of direct NEC [2] violations. Being familiar with the problems associated with incorrect installations of sensitive electronic equipment and NEC violations, all computer installations were closely inspected.

Since it is not unusual for many sections of the NEC to be violated in the examples cited, and due to the fact that the NEC is revised and the section numbering may change, exact sections are not cited.

Problem 1: Automated Cash Register and Checkout. A large food market had installed an automatic scanning checkout counter and cash register. The installation specifications called for each checkout counter installation to be isolated from all other building grounds. Under each checkout counter a separate (isolated) ground rod was to be driven and used for the equipment ground for that counter.

Each checkout counter had a motor driving a conveyor. This motor was to have its ground removed and connected to the ground rod under the counter. Should a fault occur in the motor winding, which is very likely, a high-impedance ground-loop would exist and prevent the protective device from operating. The failure of the protective device to operate would place potential on the checkout counter.

Not only was this unsafe, but the NEC [2] requires all equipment and enclosures to be connected to the same "common grounding electrode." Also, the

NEC requires a "sufficiently low impedance to facilitate the operation of the circuit protective devices in the circuit."

Solution. The inspector did not issue an occupancy permit until all the isolated grounds were interconnected.

Problem 2: Insulated-Ground Receptacle. A new ten-floor office building had installed, from the basement to the top floor, a 1/4 in by 1 in copper bus bar. The bus bar was insulated. Isolated, insulated-ground receptacles were to be used for the computer installation. The receptacle ground terminal was to be connected to this bus bar. The bus bar was connected to a ground rod, which was driven through the basement floor. The object of this misguided design was to offer ground isolation from any other ground system within the building.

The NEC [2] allows "a receptacle in which the grounding terminal is purposely insulated from the receptacle mounting means" to reduce electrical noise, electromagnetic interference. However, "the receptacle grounding terminal shall be grounded by an insulated equipment grounding conductor run with the circuit conductors. This grounding conductor shall be permitted to pass through one or more panelboards without connection to the panelboard grounding terminal . . . so as to terminate directly at an equipment grounding conductor terminal of the applicable derived system or service."

Solution. An impasse developed and the building remained unoccupied until the isolated, insulated-ground receptacles were rewired according to the NEC.

7.9 Typical Life-Safety System Problems

Problem 1: False Alarms on Smoke/Fire Detector System. A 300 000 ft^2 office/data processing facility was experiencing numerous false alarms on the smoke/fire detection system. It was determined that the cause of the false alarms emanated from a local smoke detection panel in a computer room. Further investigation revealed that room temperature thermostats were connected to the local panel and used as a local alarm. When the computer room temperature exceeded 25 °C, the local panel went into alarm, causing the main building panel to go to an alarm status with all the fire bells sounding throughout the facility.

Solution. The computer room thermostats were removed from the smoke detection panel. A separate panel was constructed to monitor and control the environmental system. The local smoke detection panel was modified to a multi-zone configuration, with two separate alarm inputs required before an output alarm was sent to the main building fire alarm panel.

7.10 Typical Misapplication of Equipment Problems

Problem 1: Ambient Temperature. An energy management company was experiencing damage to microprocessors within their energy management system. Surge suppression devices were also damaged. Monitoring with a power disturbance analyzer did not reveal any electrical disturbances that could cause equipment failure. It was determined from reviewing the equipment specifications that the ambient temperature was too high for proper operation.

Solution. An improvement in ventilation level allowed the equipment to operate successfully.

7.11 References. This standard shall be used in conjunction with the following publications. When the following standards are superseded by an approved revision, the revision shall apply:

[1] ANSI C84.1-1989, American National Standard for Electric Power Systems and Equipment—Voltage Ratings (60 Hz).[22]

[2] ANSI/NFPA 70-1993, National Electrical Code.[23]

---

[22] ANSI publications are available from the Sales Department, American National Standards Institute, 11 West 42nd Street, 13th Floor, New York, NY 10036, USA.

[23] NFPA publications are available from Publications Sales, National Fire Protection Association, 1 Batterymarch Park, P.O. Box 9101, Quincy, MA 02269-9101, USA.

Chapter 8
Specification and Selection of Equipment and Materials

8.1 General. This chapter describes the many types of power-enhancement devices that accept electrical power in whatever form it is available, and modify it to improve the quality or reliability required for sensitive electronic ac equipment. These devices perform functions such as the elimination of noise, change, or stabilization of voltage, frequency, and waveform.

The power handling and performance requirements vary depending upon each application. A wide variety of power-enhancement products are available that utilize a range of technologies and provide different degrees of protection to the connected load. The requirements of the application must first be understood, and then a cost-effective solution applied using one or more of the available products.

The job of selecting the appropriate power-enhancement device is fairly straightforward when it powers a single load. The requirements of only one load need to be considered. For larger systems that support many loads, the requirements of all loads need to be considered, as well as the potential interactions between them to decide the appropriate enhancement equipment and system construction.

Prior to addressing the selection of power-enhancement equipment, the following should be considered:

(1) *Is power quality really a problem?* Poor power quality is only one of many reasons for operational problems with critical loads. Examples of other problems that could interfere with proper operation of a critical load include software and hardware troubles within the system, temperature, and humidity beyond the limits of the critical load, electrostatic discharge, improper wiring and grounding, and operator errors. The power quality requirements of the load must be known. Refer to Chapter 3 for several guidelines.

(2) *What type of power disturbances are occurring?* To determine what type of conditioning is required, refer to Chapter 7 for guidelines on site power analysis. In addition to the current quality of power, some anticipation of the future quality and reliability of the power supply should be considered.

(3) *What level of expenditure is justified to eliminate or mitigate the power-related problems?* Some estimate should be made of the costs associated with power disturbances. This includes the value of hardware damage, lost data, lost productivity, and processing errors.

8.2 Commonly Used Power-Enhancement Devices. Table 8-1 gives a summary description of the most commonly used power-enhancement devices, and Fig 8-1 shows a summary of performance features of the various types of power conditioning equipment.

Table 8-1
Summary of Power-Enhancement Devices

Device and Principal Functions	General Description
Isolation Transformers Attenuates common-mode disturbances on the power supply conductors. Provides a local ground reference point. With taps, allows compensation of steady-state voltage drop-in feeders.	Transformer with physically different winding for primary and secondary. Often has single or multiple electrostatic shields to further reduce common-mode noise.
Noise Filters Common or transverse-mode noise reduction with attenuation and bandwidth varying with filter design.	Series inductors with parallel capacitors. Good for low-energy, high-frequency noise.
Harmonic Filters Reduction in input current harmonics of n ar loads, which can cause heating of power conductors, transformers, motors, etc.	Series inductors with harmonic trap to prevent harmonics from being fed back to line.
Surge Suppressors Divert or clamp surges.	Various types of surge suppressors are available to limit circuit voltages. Devices vary by clamping, voltage, and energy handling ability. Typical devices are "crowbar" types like air gaps, gas discharge tubes; and nonlinear resistive types like thyrite valves, avalanche diodes, and metal oxide varistors. Also available are active suppressors that are able to clamp, or limit, surges regardless of where on the power sinewave the surges occur. These devices do not significantly affect energy consumption.
Voltage Regulators Provide a relatively constant steady-state output voltage level for a range of input voltages.	A variety of voltage regulation techniques are utilized. Common techniques include ferroresonant transformers, electronic tap switching transformers, and saturable reactor regulators.
Power Line Conditioners Most often a product providing both regulation and noise reduction. Some products provide multiple noise-reduction methods, e.g., transformer and filter, but no regulation.	Shielded ferroresonant transformers or shielded transformers with tap changers, including surge suppressors and filters.

Table 8-1 *(continued)*
Summary of Power-Enhancement Devices

Device and Principal Functions	General Description
Magnetic Synthesizer Voltage regulation, common- and transverse-mode noise and surge attenuation and correction of voltage distortion.	Three-phase, ferroresonant-based device that generates an output voltage by combining pulses of multiple saturating transformers to form a stepped waveform.
Motor Generators Voltage regulation, noise/surge elimination, and waveform correction for voltage distortion.	Most often two separate devices, a motor and an alternator (generator), interconnected by a shaft or other mechanical means.
Standby Power Systems (Battery-Inverter Type) Inverter and battery backup, operating as UPS, when normal power fails. In standby mode, the load is fed directly from the utility.	An inverter to which the load is switched after utility failure. There is some break in power when the transfer to and from utility power occurs. Usually comprised of a solid-state inverter, battery, and small battery charger.
Uninterruptible Power Supplies Maintain supply of regulated voltage, waveshaping, noise/surge violation for a period of time after power failure.	Line interactive or rectifier/inverter technologies are most common. A battery supplies the power to the inverter during loss of input power.

8.2.1 Isolation Transformers. Isolation transformers are one of the most widely used power-enhancement devices. Fig 8-2 depicts the configuration of an isolation transformer. They incorporate separate primary (or input) and secondary (or output) windings. They provide for several functions. One is the ability to transform or change the input-to-output voltage level and/or to compensate for high- or low-site voltage. Typically, 480 V is distributed to the point of use and then transformed to 120 or 120/208 V. Another function of the separate windings is to provide for establishing the power ground reference close to the point of use. This greatly reduces the problem of common-mode noise induced through "ground loops" or multiple-current paths in the ground circuit upstream of the established reference ground point (see Chapter 4). These passive devices introduce minimal current distortion onto the input source. In addition, they can reduce the harmonic currents fed back to the source by three-phase nonlinear loads. When a delta primary, wye secondary, isolation transformer is used to power a load such as a rectifier, the triplen harmonic currents circulate in the delta primary so they are not seen by the power source (utility).

For power conditioning purposes, isolation transformers should be equipped with electrostatic (Faraday) shields between the primary and secondary windings as shown in Fig 8-3. An electrostatic shield is a conducting sheet of nonmagnetic material (copper or aluminum) connected to ground that reduces the effect of interwinding capacitive coupling between primary and

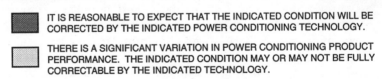

Fig 8-1
Summary of Performance Features for Various Types of
Power Conditioning Equipment

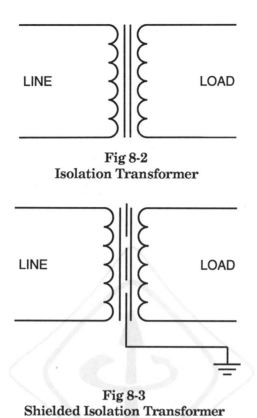

Fig 8-2
Isolation Transformer

Fig 8-3
Shielded Isolation Transformer

secondary windings and improves the isolation transformer's ability to isolate its load from the common-mode noise present on the input power source. Simple shielding adds little to the cost, size, or weight of the transformer.

Specialty conditioning transformers, referred to as "super isolation" or "ultra isolation" transformers, are equipped with additional shields around each winding to further reduce the capacitive coupling. This type of transformer is claimed to reduce the common-mode noise on the utility by 140 dB or more. However, this is done at the expense of introducing additional transformer reactance with resultant degraded voltage regulation with load change and higher costs than that of the isolation transformers with single electrostatic shields. These transformers generally do not provide decoupling of the transverse-mode disturbances.

Isolation transformers do not provide any line voltage regulation and, in fact, will cause some additional degradation of voltage regulation due to their series impedance. As was stated above, shielding tends to adversely affect regulation. Isolation transformers tend to be quite efficient (95–98%) so they generate little heat and are relatively quiet. They can be obtained in enclosures that are suitable for installation in computer rooms.

Isolation transformers can be installed separately or with power distribution breakers. Isolation transformers with distribution breakers have the advantage of being able to be located very close to the critical load. This configuration provides for short power cables that limit the amount of noise that can be coupled into them. Added advantages of these units are additional noise and surge suppression, integral distribution, monitoring, and flexible output cables that provide for simpler rearrangement of the data center.

8.2.2 Noise Filters. Line filters have the function of reducing conducted electromagnetic interference, and radio frequency interference. Fig 8-4 shows a representation of one type of LC filter. Filters can be used to prevent interference from traveling into equipment from the power source as well as prevent equipment that generates interference from feeding it back into the power line. Most types of sensitive electronic equipment have some form of filters to limit the high-frequency noise.

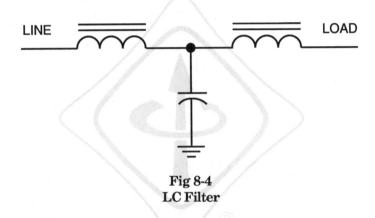

Fig 8-4
LC Filter

The simplest form of filter is a "low pass" filter designed to pass 60 Hz voltage but to block the very high frequencies or steep wavefront surges. These devices contain series inductors followed by capacitors to ground. The inductor forms a low-impedance path for the 60 Hz utility power, but a high-impedance path to the high-frequency noise. The capacitor conducts the remaining high-frequency noise to ground before it reaches the load. RFI filters are not effective for frequencies near 60 Hz, such as low-order harmonics.

Filters can be connected line-to-line or line-to-neutral for rejection of transverse-mode noise. They can also be connected line-to-neutral and line-to-ground or used in conjunction with a balun transformer to reduce common-mode noise between any of the conductors. Filters require some knowledge to properly apply them. If not used properly, they can cause a ringing effect that can be worse than the noise they were intended to filter out. For this and other reasons, filters larger than simple radio frequency interference (RFI) filters are seldom used as add-on line-conditioning devices.

8.2.3 Harmonic Current Filters. Harmonic current filters are used to prevent the harmonics of nonlinear electronic loads from being fed back into the power service where they cause heating of conductors and transformers and corresponding voltage distortion. These devices can be small units for plug-connected loads or larger devices for hard-wired loads. One variation of this type of filter is shown in Fig 8-5. The filter is placed in series with the load and the trap tuned for the predominant harmonic supplies the harmonic currents required by the load. These filters can be very effective at greatly reducing the harmonic currents at their source and eliminating the need for other changes to compensate for the problems caused by the harmonic currents.

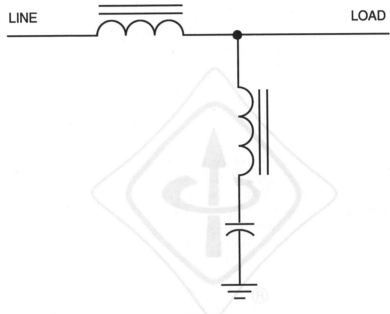

Fig 8-5
Harmonic Current Filter

8.2.4 Surge Suppressors. Surge suppressors encompass a broad category from large devices, such as lightning-surge arrestors to small suppressors used to protect plug-connected devices. Effective surge protection requires the coordinated use of large-capacity current-diverting devices at the service entrance followed by progressively lower voltage-clamping devices. The service entrance devices are intended to lower the energy level of a very large surge to that which can be handled by other devices closer to the loads. If improperly coordinated, excess energy can destroy the downstream suppressors and damage the connected load equipment.

The smaller surge suppressors are generally simple, and relatively low-cost, devices. They usually contain metal oxide varistors, avalanche diodes, or other voltage-clamping devices that are connected across the power line or from one phase-voltage lead to another or to ground. Suppressors absorb or divert energy from surges that exceed their voltage threshold (typically 100%

above the nominal line voltage). Because of their small size and low cost as compared with the equipment they serve and the cost of determining if such surges exist at a given installation (or even if this feature is already built into the computer itself), they are often routinely used as low-cost insurance against the chance of severe surges. Many of the higher quality line conditioners include suppressors. They can even be added to a distribution panelboard if not included elsewhere.

Surge suppression devices are packaged into various assemblies that often include power receptacles for several loads. These units are most commonly sold for use with small, single-phase loads and are available from a variety of manufacturers. The better units include fusing, agency listing, and surge capability in the form of a current rating. Most of the lower-cost units have limited ability to survive and to protect the load against large surges. The protective device may fail without any indication that the unit is unable to function.

8.2.5 Voltage Regulators. Most low-frequency disturbances, except outages, can be handled by appropriate application of a voltage regulator. There are a number of types of voltage regulators in use today. Solid-state devices, such as constant voltage and tap-changing transformers, are used almost exclusively today, rather than slower acting electromechanical types.

Early electromechanical regulators typically had a motor operator that moved a sliding tap on a transformer. These induction-regulators are fast enough to follow voltage changes that occur during the day or seasonally due to application and removal of steady load. These units are not suitable to protect sensitive electronic load against rapid changes in voltage.

8.2.5.1 Tap Changers. Fast response regulators divide into two generic classes, tap changers, and buck-boost. The first is the tap changing regulator shown in Fig 8-6. Quality tap changers are designed to adjust for varying input voltages by automatically transferring taps on a power transformer (either isolating type or auto-transformer type) at the zero current point of the output wave. Some models make the tap change at the voltage zero crossing, which causes a transient to be generated except when the load is at unity power factor. With voltage-switching-type units, the magnitude of the transient should be determined on the actual load conditions. The number of taps determines the magnitude of the steps and the range of regulation possible. A good quality regulator will have at least 4 taps below normal and 2 taps above normal for 7 total steps. The taps are usually around 4–10% steps, depending on specific designs. Response time is usually 1 to 2 cycles and is limited to that speed because of the zero current switching criteria. Practical sense time and control system stability typically limits full correction time to 3–5 cycles.

A major advantage of the tap changer is that its only impedance is the transformer or auto-transformer and the semiconductor switches. It introduces little harmonic distortion under steady-state operation and minimizes load-induced disturbances as compared to regulators with higher series impedance. It also has high short-term overload capability that will provide for starting inrush. In its usual configuration with an isolating transformer and wide undervoltage capability, it provides both common-mode isolation and regulation.

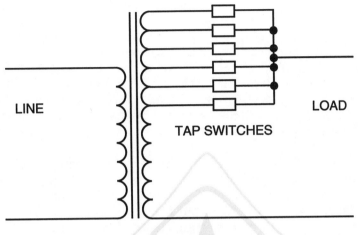

Fig 8-6
Tap Changing Regulator

8.2.5.2 Buck-Boost. The second class of fast response regulators is the electronic buck-boost type (Fig 8-7). It utilizes thyristor control of buck and boost transformers in combination with parametric filters to provide regulated

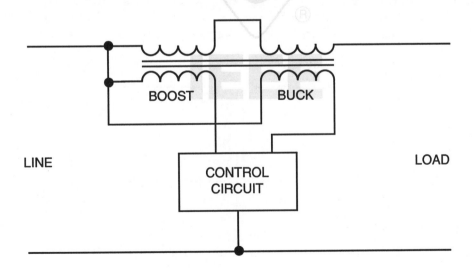

Fig 8-7
Buck-Boost Regulator

sinusoidal output even with nonlinear loads typical of computer systems. This is done in a smooth continuous manner eliminating the steps inherent in the tap changer. Inrush currents can be delivered for start-up typical of computer central processors or disc drive motors while maintaining nearly full voltage. Units can be equipped with an input isolating transformer with electrostatic shield providing voltage stepdown and common-mode attenuation when needed. Power is fed to the regulator, which either adds to (boosts) or subtracts from (bucks) the incoming voltage so that the output is maintained constant for 15–20% variations of input voltage. This is done by comparing the output voltage to the desired (set) level and by the use of feedback to modify the level of boost or buck so that the desired level is maintained. A parametric filter provides a path for nonlinear currents generated by the load and by the regulator itself and produces a sine wave output with low total harmonic distortion.

8.2.5.3 Constant Voltage Transformers. One common type of regulator is a "ferroresonant" or constant voltage transformer (CVT). Fig 8-8 represents one design topology of a ferroresonant regulator. This class of regulators uses a saturating transformer with a resonant circuit made up of the transformer's inductance and a capacitor. The regulator maintains a nearly constant voltage on the output for input voltage swings of 20–40%. These units are reliable because they contain no moving or active electronic parts. If these units are built with isolation (and shielding), they can provide for common-mode noise reduction and provide a separately derived source for local power grounding. They also attenuate normal-mode noise and surges.

Careful analysis is required for overload conditions. The load current tends to cause the unit to go out of resonance if it gets too high. Often these units can only supply 125–200% of their full load rating. If inrush or starting currents exceed these limits, the output voltage will be significantly reduced, which may not be compatible with many loads. The other devices on the output of the CVT will see this sag in the voltage and may shut down due to an

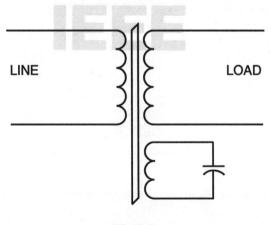

**Fig 8-8
Ferroresonant Regulator**

undervoltage. These devices should be oversized if they are expected to provide for heavy starting or inrush currents.

Constant voltage transformers draw current all of the time. This current is due to the resonant circuit and causes these units to be less efficient at light loads, as compared to other types of regulators. Some of the units are quite noisy and require special enclosures before they can be installed in office environments. For more information on ferroresonant or constant voltage transformers, see IEEE Std 449-1990 [B3].

8.2.6 Power Line Conditioners. Typical power line conditioners combine one or more of the basic power-enhancement technologies to provide more complete protection from power disturbances. Some line conditioners combine the noise-reduction features of isolation transformers or filtering devices with voltage regulators. These units provide a locally derived source with isolation while providing voltage regulation. The advanced conditioners also incorporate surge suppressors to clamp high-voltage surges, which filtering alone does not address.

8.2.6.1 Magnetic Synthesizer. Another ferroresonant-based technology is the magnetic synthesizer (Fig 8-9). These units consist of nonlinear inductors and capacitors in a parallel resonant circuit with six saturating pulse transformers. These units draw power from the source and generate their output voltage waveform by combining the pulses of the saturating transformers in a stepped-wave manner. They provide for noise and surge rejection and regulation of the output voltage to within 10% over large swings in input, up to plus or minus 50%. These units generally incorporate shielding into the pulse transformers to attenuate common-mode disturbances. Additional filtering is included to eliminate self-induced harmonics. This filtering can handle a reasonable level of harmonic distortion at the input or at the output as induced by the nonlinear loads. The circuit is tuned to the rated output voltage and frequency.

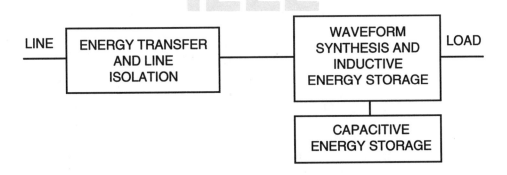

**Fig 8-9
Magnetic Synthesizer**

The regulator has an inherent current limiting characteristic that limits maximum current at full voltage to the range of 150–200% of rating. Beyond that load, the voltage drops off rapidly producing typically 200–300% current at short circuit. This is a limitation with large inrush and starting currents. Sudden large load changes, even within the units rating, can cause significant voltage and frequency transients in the output of this type of line conditioner. These units are best applied when the load does not make large step changes.

The tuned circuit has stored energy and will, therefore, ride through outages of one-half cycle or slightly more provided the outage is not a fault close to the input, which would drain the stored energy. Magnetic synthesizers tend to be large and heavy due to the magnetics involved and can be acoustically noisy without special packaging. Some of the larger units display good efficiencies as long as they are operated at close to full load. Depending upon the design, the synthesizer may introduce some current distortion on its input, due to its nonlinear elements.

8.2.6.2 Motor-Generators. Motor-ac generator sets (M-Gs) provide the function of a line conditioner and can also provide for conversion of the input frequency to a different frequency that is required by the load. Fig 8-10 depicts one configuration of a M-G. Examples of this are 60–50 Hz or 60–415 Hz frequency converters. These units consist of a utility-powered electric motor driving an ac generator that supplies voltage to the load. The motor and generator are coupled by a shaft or belts. This totally mechanical coupling of the input and the output allows the M-G to provide total noise isolation of the load from the input power source. Practical M-G systems include a bypass circuit that can reduce this total input-to-output isolation.

Fig 8-10
Motor-Generator Set

Shaft or belt isolated M-Gs are used widely as a source of 415 Hz power for large computers requiring this frequency. Because the frequency tolerance is wide, a single 60 Hz induction motor can be used to drive a brushless synchronous 420 Hz generator or alternator. The induction motor is the least expensive of the common types of motors used on these devices. This type of motor does not rotate at the same speed as the rotating field that is generated by the input power. The speed at which the motor turns changes with load and input voltage variations. Since the generator frequency is a function of its shaft speed, the output frequency varies with the motor speed. The output voltage is maintained by controlling the excitation to the field winding of the generator and is independent of small changes of motor speed.

For a 60 Hz output, however, the frequency tolerance of the load can be much more critical. Computers generally require plus or minus 0.5 Hz. This can be achieved by the use of low-slip designs in the induction motor. In the most critical applications, a synchronous motor drive is required so that output frequency is the same as the utility input. Totally synchronous M-Gs also maintain their outputs nearly in phase with the utility source. This allows for uninterrupted transfers between the M-G and utility for maintenance.

M-Gs protect the load from voltage sags, swells, and surges. For power line voltage changes of plus or minus 20–50%, voltage to the load is still maintained at nominal. A useful feature of the M-G is its ability to bridge severe short-term sags or outages. The rotational momentum of the rotating elements permits the M-G to span momentary outages of up to about 0.3 s. The M-G ride-through time may be affected if the power outage is some distance from the conditioner so that it appears as a short on the input by virtue of other loads connected to the same source. Part of the rotating energy stored in the M-G can be lost by the dynamic breaking action of the motor. The limiting factor is the drop in frequency shaft speed that can be tolerated as energy is removed from the M-G set. This period can be extended by adding inertia via a flywheel. Ride-through times of several seconds are available through the use of large flywheels.

Products are available that are able to maintain output frequency even while the shaft speed is slowing down. These devices do not have fixed poles in the generator. Instead, the poles are created or "written" as the device rotates. When input power is lost and the shaft speed starts to decay, the spacing of the poles is reduced and their number is increased so that the frequency remains constant. This method achieves ride-through times that are significantly longer than other devices with the same rotating energy at the cost of increased complexity.

Another form of M-G is referred to as a rotating transformer. These units have a common rotor with two stators. One is the motor stator and the other is the generator's stator. These are compact units that have demonstrated excellent efficiency. One drawback of this design is that they do not provide the same level of noise and surge isolation between the input and the output as conventional M-Gs. The noise has a path through the unit because of the coupling between the two stators that are typically wound one on top of the other.

M-Gs tend to be more expensive than other types of line conditioning equipment. They are usually physically large and heavy. Depending on the design, the M-G efficiency can be relatively low so that electrical energy costs over its lifetime may be substantial. The rotary transformer types, as well as some of the larger standard units, display better performance in this area. M-Gs tend to be noisy and require soundproof enclosures to make them suitable for computer room installation. M-Gs do not introduce measurable current distortion on their input source and have the added advantage of lowering the overall level of distortion by isolating the utiity from the harmonic current requirements of the loads supplied by the generator.

8.2.7 Computer Power Centers. A computer power center is a device that provides a convenient method for distributing electrical power to many devices without the need for hard wiring, and forms a separately derived source for local grounding. It is essentially a cabinet with a flexible input cable, isolation transformer, distribution circuit breakers, and flexible load cables. The load cables are terminated with mating connectors for direct connection to the load equipment. Some manufacturers include power conditioners such as tap-changers, M-Gs, and synthesizers internal to the power center to further enhance performance.

The computer power center greatly reduces the time required to install the average computer system and allows for relatively easy relocation of equipment as compared to hard wiring methods. This can translate into significant cost and time savings. The isolation provided by the transformer (or M-G) in the power center allows the creation of a local ground as described in FIPS PUB 94 [B1][24] to greatly improve system reliability.

Units with internal power conditioners can be used to reduce the effects of long distribution lines from central power conditioning and uninterruptible power supply (UPS) equipment. The effect of current harmonics introduced on the power source is a function of the type of conditioner used in the power center.

8.2.8 Standby Power Systems (Battery-Inverter Type). Standby power systems are those power systems in which the load is normally supplied by the utility input. Fig 8-11 shows one configuration of the standby system. The standby system only supplies the load when a satisfactory utility source is not available (IEEE Std 446-1987 [B2]). These power systems are intended for loads that can tolerate the momentary loss of power during the transfer. They come in a number of configurations using a number of technologies, and are used for a variety of loads ranging from personal computers to emergency lighting.

The simplest form of standby system has the load connected to the utility source through a transfer switch during normal operation. In the event of a utility failure, the load is transferred to an inverter that generates ac power of satisfactory quality to support the load. The inverter is fed from a battery that has been maintained at full charge from a rectifier unit that is fed from the utility source. The design of this type of supply allows several economies. First, the inverter is not supporting the load on a continuous basis. It only has to operate for the duration of the power outage or for the life of the batteries. This period is typically 15 min or less. Second, the rectifier section only has to be able to recharge the battery and not support the full load of the inverter.

Normal operating efficiency of this type of unit is high since the load is being fed from the utility under normal operation. The losses are those associ-

---

[24] The numbers in brackets correspond to those in the bibliography in 8.7.

ated with the line conditioning element (if used), rectifier, and the inverter, if in fact it is operating while off line. A major requirement of this type of unit is its ability to sense all types of power failures and transfer to the inverter without an unacceptably long input-power loss to the load equipment. These units are typically successful in powering systems that have power supplies that can tolerate short durations of input-power interruption. They are often employed with loads that utilize switch-mode power supplies, which often do not require regulated input voltage and are tolerant of momentary loss of power during the transfer. In addition, fast electronic (static) transfer switch can be used to greatly reduce the transition time.

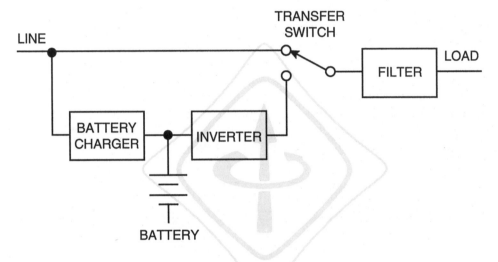

Fig 8-11
Standby Power System

The next level of sophistication involves the use of a line conditioner in series with the load to provide conditioning of the utility voltage during normal operation. The conditioner can be one of the types that were previously discussed. Some manufacturers take advantage of the extensive filtering capability of some of the conditioners, such as the ferroresonant transformer, magnetic synthesizer, and M-G. The filtering capability allows them to use a very simple inverter circuit that generates square waves as opposed to sine waves. The line conditioner is in circuit all the time and provides conditioning of the inverter output as well as the utility during normal operation. Continuous regulated output power can be achieved by this method if the line conditioner has sufficient ride-through to power the load during the interruption time (see 8.2.9.2).

Another variation of this topology is one that has a tap changer that powers the load, and an inverter that, under normal operation, is used as a rectifier to maintain the charge of the battery. When the input power source fails, the inverter shifts phase quickly to start taking power from the battery and supplying the load through the tap-changing transformer.

8.2.9 Uninterruptible Power Supplies (UPSs). UPSs are intended to provide regulated output power regardless of the condition of the input power source, including total power outages. UPSs come in a variety of configurations and utilize various technologies. The major categories of UPS are rotary and static UPS.

8.2.9.1 Rotary UPS. A rotary, or M-G UPS, consists of a rotary line conditioner modified to receive power from a battery when utility power is not available. Three major methods are used to provide this uninterruptible performance.

One method involves the addition of a dc motor to the system (Fig 8-12). The dc motor takes over for the ac motor when the utility power is no longer sufficient to support the load. These motors can be on the same shaft as the rest of the M-G or can be connected by drive belts. The battery can be recharged by a solid-state battery charger or can be recharged directly from the dc motor. This is accomplished by controlling the field current to change the function of the dc motor to that of a generator. This approach reduces the complexity of the system but the dc motor usually experiences rapid brush wear when operated in this idle state.

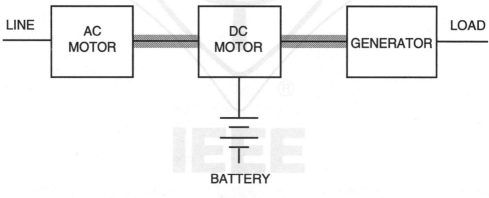

Fig 8-12
Rotary UPS With DC Motor/Generator

Another method involves a M-G with a dc motor driving the alternator (Fig 8-13). The dc for the motor is derived from a solid-state rectifier which also charges the system batteries. The one-line diagram of this supply looks very much the same as a solid-state UPS, except the solid-state inverter has been replaced with a rotary inverter.

The other common method involves the use of a static inverter/motor drive to supply ac power to the motor during utility power outages (Fig 8-14). When utility power is lost, the inverter converts the power from the batteries into

60 Hz ac which is supplied to the input of the motor. This switchover is accomplished during the ride-through time that the inertia of the M-G provides. The inverter can be used as a battery charger during the time that the utility ac is available to charge the battery. Separate battery chargers are also used to perform this function.

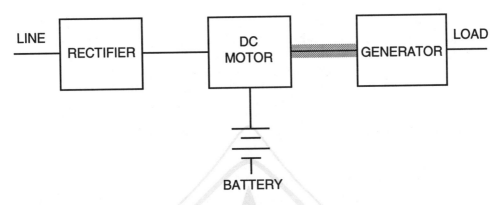

Fig 8-13
Rotary UPS With Rectifier/DC Motor

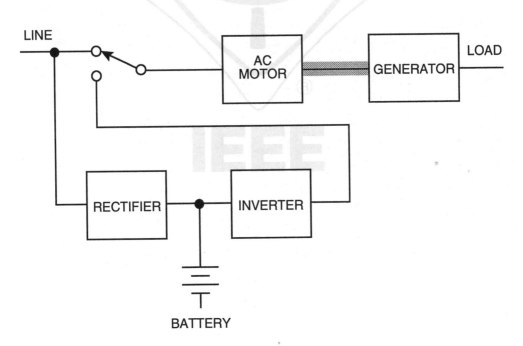

Fig 8-14
Rotary UPS With Inverter

A bypass transfer switch is usually included with a rotary UPS. These switches can be of the solid-state type or strictly mechanical. The switch transfers the load to the utility for maintenance of the UPS or during a failure within the UPS itself. With synchronous M-Gs, the transfer can be made at any time due to the inherent near phase lock of the output with the utility source.

The amount of current distortion introduced by the rotary UPS is a function of its design. Units without a solid-state rectifier do not introduce harmonics on the source and actually can reduce those already there due to other nonlinear loads sharing the same input service. Units with rectifiers that are used only to charge the batteries will typically introduce significant current distortion during battery recharging with only a small amount after battery has been recharged. A rotary UPS that has a rectifier supplying a dc motor will introduce current distortion based on the type of rectifier and amount of filtering provided. These units are equivalent to static UPSs that utilize similar rectifier sections.

8.2.9.2 Static UPS. The static UPS is a solid-state device that provides regulated continuous power to the critical loads. Static UPSs fall into two basic designs: rectifier/charger, diagrammed in Fig 8-15(a), and line-interactive, diagrammed in Fig 8-15(b).

In the rectifier/charger (or double conversion) UPS, input power is first converted to dc. The dc is used to charge the batteries and to constantly operate the inverter at full load. In the line-interactive (or single conversion) UPS, utility power is not converted into dc but is fed directly to the critical load through an inductor or transformer. Regulation and continuous power to the critical load is achieved through the use of inverter switching elements in combination with inverter magnetic components, such as inductors, linear

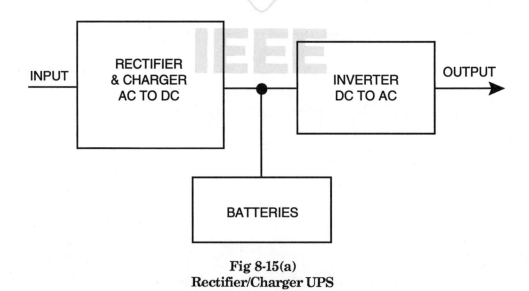

Fig 8-15(a)
Rectifier/Charger UPS

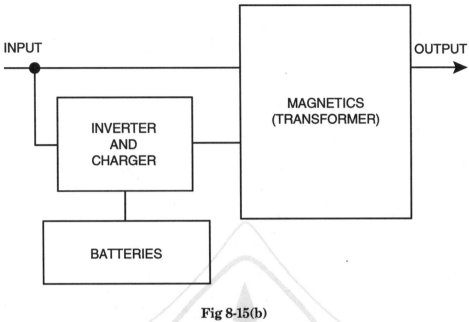

**Fig 8-15(b)
Line Interactive UPS**

transformers, or ferroresonant transformers. Some systems have relatively complicated controls on the inverter and are able to provide improved output voltage regulation. In this case the inverter controls its phasing and duty-cycle to both charge the battery and provide a voltage component to be summed in the transformer. The term "line interactive UPS" comes from the fact that the inverter interacts with the ac line to buck, boost, or replace incoming ac power as needed to maintain voltage control.

8.3 Equipment Procurement Specifications. Generation of the specification for the required power-enhancement product is a very important part of the procurement of the system. There is a large number of different specification items that are published by the manufacturers. Some of the items are of universal importance to all users and some are of more interest in one application than another. The procurement specification should emphasize those specifications of particular interest for the application. Any items that can have the specification loosened should be treated appropriately in the procurement. In this way the specification defines the system requirements without over-specifying. This approach helps assure the procured products are the best combination of performance and price for the requirements of the particular installation.

The more important product specifications should be discussed in groups, organized according to their area of importance. Specification items are grouped into the categories of installation considerations, reliability considerations, and cost of operation.

8.3.1 Facility Planner's Considerations. The following items are in the areas that are of interest to the installation planner. They encompass those items that describe the primary operation of the system. A description of each is given.

8.3.1.1 System Load Rating. This rating is the basic capacity rating of the system. It is expressed in both apparent power, kVA, and power, kW. The apparent power, kVA, is simply the maximum current that the system can support times the line-to-line output voltage (times the square root of 3 for three-phase systems). For most loads, the actual power used, kW, is less than the kVA value. The ratio of the kW to the kVA is called the power factor. In the case of linear loads, the power factor is a function of the phase shift between the applied voltage and the current that the load draws. In the case of a nonlinear load, the power factor is a function of this same phase shift as well as how much the current waveform varies from a pure sine wave. The power factor rating of any power conditioners should take into account the portion due to phase shift and the portion due to waveform distortion. The crest factor rating discussed below should be used to determine the conditioner's ability to support loads that have high levels of input current distortion. If the load power factor is anything but 1.0, the actual power, kW, that the system is supplying will be different. Most systems are rated at a power factor between 0.7 and 0.9. This means that the actual power rating of the system will be less than the kVA rating. Both ratings are important because neither can generally be exceeded at steady-state conditions.

There are several factors that may require that the rating of the power conditioner exceeds the steady-state load requirements. First, many loads require more current during starting than they do under normal operation. In a similar manner, some loads have periodic increased load requirements that should be taken into account when sizing the power conditioning system. In addition, the potential growth requirements in the near future should be considered. Load requirements typically grow with time, and various economies can be achieved if this growth is anticipated and accommodations are made during the initial planning.

The rating of the power conditioner may also vary with the type of load that is applied. Many modern loads have rectifiers or switching inputs that do not draw current in a smooth manner at the input power frequency. This current distortion can cause additional stress on the power conditioner circuits that in turn will affect the rating of the conditioner when supplying these loads. The conditioner manufacturer should specify the rating conditions. The ability to support these nonlinear loads is usually stated as a "crest factor" that describes how much the load current can vary from a pure sine wave while maintaining the system's full rating. In the case of crest factor, a linear load has a factor of 1.414, which is the ratio of the peak value of a sine wave to its rms value. Therefore, a load with a crest factor rating of 2.8 is twice the peak current requirements of a linear load, not nearly three times.

8.3.1.2 Size and Weight. The size of the system is important because of the cost or lack of floor space that is available for the system. The weight is important because of floor loading limitations as well as elevator ratings.

8.3.1.3 Air Conditioning Requirements. These requirements are a function of the efficiency of the system and must be considered when sizing the air conditioning system for the installation. The heat loss is generally specified in British thermal units per hour (Btu/h). Also included is the recommended operating temperature and humidity range that determines the kind of air conditioning or ventilation system that will be required.

8.3.1.4 Audible Noise. The amount of noise that is generated varies greatly from one system to another. The noise level is of great importance if the system is to be installed in the computer room or in, or close to, offices. Many systems are available with additional soundproofing or special enclosures to reduce the sound level emitted.

8.3.1.5 Battery Configuration and Life. There are two main types of batteries (wet-cell and maintenance-free) used for UPS applications. The original type is the wet-cell battery. This type is used in large installations with long back-up times. Wet-cell batteries are generally installed on open racks, usually in their own enclosed room with separate ventilation from the rest of the facility. Ventilation is required because, under certain conditions, the batteries generate hydrogen gas. Often hydrogen detectors, temperature detectors, showers, and eye washes are required by local code. All of these items add to the cost of the installation. Some of the considerations discussed below may become more or less significant as battery technology evolves.

The life of the wet-cell battery is affected by the environment and the operating conditions. Most battery manufacturers specify that the average temperature in the battery room should be 25 °C. At low temperatures, the battery capacity (back-up time) is less than normal. The battery capacity and loss of electrolyte increases as the temperature increases. These batteries generally have a specified life of 10 to 20 years. The rate of internal breakdown within the battery increases with temperature. The effective life of the battery can be significantly shortened by operating at elevated temperatures. Battery life is also a function of the number of discharges and the depth of discharge. Wet-cell batteries in UPS applications can have a useful life on the order of hundreds of discharges.

In recent years, sealed maintenance-free batteries have been used in increasing numbers for UPS applications. These units can be housed in cabinets or placed on open racks. They require minimal maintenance during their life. These batteries do not generate significant gas during normal operation. Their low-gassing level allows the battery cabinets to be installed almost anywhere, including on the computer room floor next to the UPS. The special requirements for wet-cell batteries are generally not required. If the batteries are located right next to the UPS cabinet, the amount of cabling required is greatly reduced. All these items generally make the maintenance-free batteries much less expensive to install.

Depending upon design and mission objectives, the rates life of batteries can range from 2 to 20 years. Their actual life is affected by the same conditions as the wet-cell batteries. By definition, maintenance-free batteries have a limited amount of electrolyte, which is not replenished during their life. This fact can make them more sensitive to operation at elevated temperatures and high ripple currents (see IEEE Std 446-1987 [B2]).

In most operations the batteries are maintained at what is called their float-voltage. This is the voltage that will allow the batteries to become fully charged but not overcharged. The battery accepts the amount of charge necessary to maintain full charge and no more. Most UPS batteries are made up of cells that are connected in series to achieve the desired voltage level. Since these batteries are wired in series, the same current must flow in each battery. If one battery tends to self- discharge a little faster than the rest, it will slowly become less charged than the rest. This situation is detected by periodically measuring the voltage across each battery to verify that they are closely balanced. If the voltages vary beyond limits, an equalizing charge is performed. This charge involves raising the charge voltage above the float value for a specified length of time. This charge-cycle forces additional charge current to flow through all of the batteries. The lower voltage cells are brought up to full charge and the others are slightly overcharged. It is often necessary to parallel strings of batteries to achieve the desired amount of back-up time.

Another operating condition common in UPS installations that increases the rate of breakdown is ripple current. Ripple current is caused by the ripple voltage of the battery charger output and by the pulsating current requirements of the inverter. The detrimental effects of ripple current on the battery are mainly a function of the design of the charger and the relative size of the battery as compared to the UPS rating. The ripple current tends to heat the batteries and is equivalent to constantly discharging and recharging the battery a tiny amount. When high ripple current is present at elevated operating temperatures, the battery life can be reduced to one quarter of what would normally be expected.

8.3.1.6 Inrush. Inrush is the amount of current that a load draws when it is first turned on. Inrush is generally caused by the magnetization requirements of input transformers and starting requirements of fan motors and is present on most systems irrespective of soft-start feature. Inrush must be considered when sizing the electrical feed to the system.

8.3.1.7 Input Soft-Start. Input soft-start is the time that the input section (rectifier, motor, etc.) of the load requires to go from the off state to carrying the full load of the system. It is of primary importance when the system is to be powered from an engine-generator.

8.3.1.8 Input Power Factor. The input power factor of the system specifies the ratio of input kilowatt to input kilovoltampere at rated or specified voltage and load. The power factors of some conditioners are a function of the load, and some are independent of the load. Those that are a function of the load will typically be specified for a unity power factor load that does not represent normal operation. In power systems that utilize phase-controlled rectifier inputs, the input power factor will become lower or less desirable as the input voltage is raised. Other rectifier designs are becoming available that can maintain a constant or unity power factor over their full operating range.

For a given load on the power conditioner, the lower the power factor, the more input current will required by the system. The wiring to the system and the switchgear depends on the current that is drawn. All other aspects being equal, the UPS with a higher power factor over the operating range will potentially have a lower installation cost.

8.3.1.9 Input Current Distortion. The current that is drawn from the supply by most power systems contains frequency components that are harmonics of the supply frequency. These harmonic currents cause the input current to distort from a perfect sinewave. The amount of this distortion is specified as a percentage. The total harmonic distortion refers to the rms sum of all of the harmonic components. Different rectifier designs create different amounts of current distortion. This current distortion is translated into voltage distortion on the utility line in proportion to the source impedance of the utility feed. This voltage distortion can adversely affect other equipment that is powered from the same source. Lower levels of current distortion cause lower voltage distortions, and other devices are less likely to be adversely affected. Input current distortion is specified for a given set of conditions and can be affected by such factors as input voltage, load, input phase-balance, and source impedance.

8.3.2 Reliability Considerations. When one considers the purchase of a power line conditioner to protect a critical load, a primary concern should be the reliability of the system. The principal function of the system is to supply quality power to the critical load in a continuous manner. There are many items that affect the overall reliability of the system. Some of these are discussed below.

8.3.2.1 System Configuration. The reliability of system is very much influenced by its configuration. There are many options that exist to improve the basic reliability of the power converter itself. Some of these items will be discussed in the following paragraphs.

8.3.2.1.1 Parallel Systems. The most common reliability enhancement involves paralleling multiple power-conversion modules. Fig 8-16 diagrams a system with three modules. These installations are termed "redundant" since they contain at least one more unit than is required to support the load. The basis for this arrangement is that if one of the power-conversion modules fails or must be taken off line for service, the remaining units are able to support the load. This method can provide significant improvement to the system reliability. The power-conversion modules must be properly designed and installed in order to achieve this method's potential.

Parallel redundant systems can actually be less reliable if the power-conversion units are not tolerant of disturbances and overloads on their outputs. The performance of UPS equipment varies in this regard from one manufacturer to another. Methods of determining specific equipment performance are discussed in the testing section of this chapter.

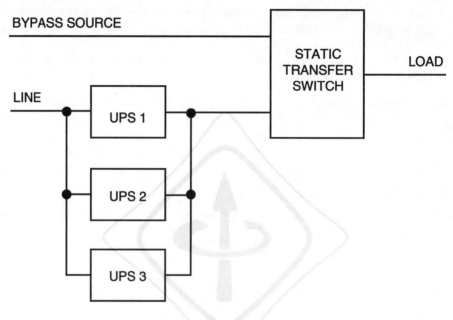

Fig 8-16
Parallel Redundant System

8.3.2.1.2 Isolated Redundant Systems. The basic configuration has two units as shown in Fig 8-17. Each unit is able to supply the full load and contains a static transfer switch. The primary unit that is connected directly to the load has its static switch fed from the secondary unit. The secondary unit's static switch is fed from the utility as are the inputs of the two units. In normal operation, the primary unit supplies the load. If it fails, the load is supplied by the secondary unit through the primary unit's static switch. If the secondary unit fails, the load is supplied from the utility through both static switches. There are other configurations of isolated redundant systems that include more than two modules. They all share the basic concept where one or more modules form a back-up source of conditioned power for one or more other modules.

This configuration does not have the same problems that paralleled systems have when a module fails. The faulted module does not have the possibility of pulling down the paralleled unit through excessive fault current. The secondary unit does see a load step equal to the existing load that will cause some transient on the load bus.

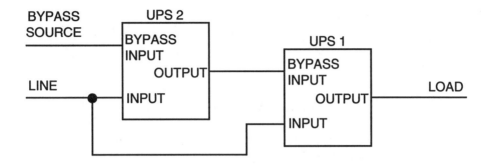

Fig 8-17
Isolated Redundant UPS

8.3.2.2 Product Reliability. Many factors are involved in making a product as reliable as possible. These factors include design, component selection, workmanship, and conservative rating of the units. It is very difficult to look at a product and determine its potential for reliable operation. Estimating the relative performance of various products usually requires sorting through reliability information from manufacturers. The following is a discussion of the items that should be considered in determining the reliability of a power line conditioner.

8.3.2.2.1 Calculated Reliability (MTBF). Most manufacturers calculate the reliability of their systems in the same manner that is prescribed for military products. The process involves determining the basic reliability of each component that goes into the system. The reliability estimate is based on field experience and accelerated life testing. The stress that is placed on the device in the application needs to be taken into consideration. Once the reliability of each component is estimated and the total number of each is known, the total system reliability can be estimated. The overall reliability of a system is usually expressed as mean time between failures (MTBF). The MTBF is usually expressed in hours, and specifies the average number of hours that can be expected between failures in the system.

The calculated MTBF is only an estimate and may not really define the actual reliability of the product. The inaccuracies come about due to the many variables that are hard to determine. Such items as the stress (peak and average current and voltages, junction temperatures) on the devices are often hard to determine accurately. Other factors associated with the design are almost impossible to estimate accurately, such as noise susceptibility, effects of accumulated dust combined with humidity, and the thoroughness and correctness of the design. Improper reactions of the system to faults or disturbances can cause the system to fail, but are not included in component reliability. Proper maintenance and installation are also often assumed in MTBF calculations. In general, calculated MTBFs should be used as guides when actual field data are not available.

8.3.2.2.2 Field Reliability Data. Once a product has been in the field for a period of time, an accurate determination of the operating reliability can be made. This procedure involves keeping track of the number of hours that the installed base of units has operated and the total number of failures that have occurred. MTBF data derived this way can be used to compare the reliability of various systems. One must make sure that the numbers from the different manufacturers are derived by the same methods. There are three reliability figures-of-merit for UPSs that should be compared: (1) individual module, (2) multi-module, and (3) total system.

First, the reliability of the individual unit or module should be examined. This figure-of-merit is a measure of how often service will be required, and should provide the means to determine the relative service costs between products.

The second figure-of-merit is the reliability of a multi-module, redundant system output. (How often did the power system itself fail to provide power that was within specifications?) This figure-of-merit can be difficult to determine unless a line-disturbance monitor is installed on the critical-load bus. Typically, the frequency of system failure (that required the critical load to be powered from the utility or alternate source) can be determined. This frequency shows how well the manufacturer's methods of providing redundancy actually performs in the field. The ratio of the module reliability and the multi-module redundant system reliability is important. Systems that have effective methods of isolating a failed module before it degrades the critical bus typically show higher ratios between the system and the module figures-of-merit. In most applications this is the most important factor because it reflects how well the load can be supported by the system, independent of the quality of the input power. This ratio is what actually justifies the purchase of a UPS.

The third reliability figure-of-merit is the most commonly stated. This figure-of-merit is the total system MTBF, including the static transfer switch that connects the critical load to the utility in the event of a complete failure of the UPS. The difference between this parameter and the previously discussed ratio will give an indication of how well the static transfer switch logic functions and how well it is integrated into the entire system. The number is somewhat dependent on the reliability of the bypass source so will vary with installation. This parameter does not distinguish between the time that the critical load is powered from the UPS output and the time that it spends on the utility source. Because of this, it is not as important in determining the reliability of the UPS itself as the ratio that does not include the alternate source and the static transfer switch.

8.3.2.2.3 Manufacturer's Experience. The field reliability data that was discussed requires that the manufacturer have a large number of products installed in the field for a long enough period of time to give an accurate picture of the products' reliability. When a new product enters the market place, the only data that exist is the calculated MTBF, which is only an estimate of what the actual reliability should be in the field. The actual (demonstrated) system reliability will not be known until sufficient field experience is obtained.

If the product is one that has been on the market for a number of years, the reliability should be easy to determine from manufacturer's data and through contacts with users of the product. It is always a good idea to contact organizations that have used the product for a number of years to see what kind of reliability and general experience they have had with the product and the manufacturer. Since reliability may vary from one application to another, it would be good to talk to several organizations that have similar quality of input power and similar equipment on their critical bus.

If the product is new to the marketplace, the user will have to rely on calculated reliability data and very thorough testing of the product before it leaves the factory as well as after it is installed on site. The manufacturer should demonstrate control of the product configuration and its production processes. A comprehensive quality assurance program should be in place. A purchaser of a new product should review the technical aspects of the manufacturer's operation and be convinced that quality control really exists.

It is difficult to totally test a power-enhancement system in a factory situation. The best that can be done is to test each of the key performance features of the system as completely as possible. After the system successfully passes all the tests, it should undergo an extensive burn-in at the factory and on site before the system is placed in service with the critical load. The burn-in should be completed without failure of the critical bus.

8.3.3 Installation Cost Considerations. There are a number of factors that affect the final cost of installing a power-line conditioner or UPS. These costs should be considered along with the purchase price for each of the possible systems that are under consideration. Some of the factors that can affect the installation costs are discussed below.

8.3.3.1 Location of the Installation. There are several options as to where a power-line conditioner system can be installed. Installation of smaller systems tends to be rather straightforward and the costs involved are usually not very great. Small UPS typically have sealed maintenance-free batteries that are installed inside the cabinet or can be placed right next to the UPS, which further simplifies the installation. The smaller systems generally feed a limited number of loads so that distribution is less of a problem. The very small units typically have power receptacles into which the loads can be directly connected.

The installation of larger systems is different. These systems tend to be large and generate significant heat and noise. Very large UPS systems have stationary wet-cell type batteries that require that numerous safety precautions be taken, resulting in very significant installation costs. Most of these large systems have been installed in special rooms designed for the UPS or have been placed in existing equipment rooms.

Many of the factors that affect the cost of installation are based on the physical constraints that are placed on the installation by the available space. There are certain items that can somewhat be controlled to lower the cost of installation. The most common ones are described below.

A new generation of UPS products is emerging that allow alternative installation options. With these products, it is possible to install the entire system on the computer room floor, offering some real advantages. The floor space in the computer room tends to be more expensive than other installation sites. These new products reduce the floor space required due to their compact designs.

In addition, the efficiency of these systems tends to improve with each new generation. Since the heat loss is a function of the system's efficiency, the newer products tend to dissipate less heat than older ones. More emphasis is being placed on the noise level emitted by the system. The noise level is being reduced by use of baffling, newer fan designs, and switching techniques that place most of the noise above the audible range. These factors make these power products easier to place in areas that are also working areas for personnel.

8.3.3.2 Wire and Breaker Costs. The cost of the electrical cabling and breakers is a function of the current that the system draws and supplies to the loads. The efficiency of the system as well as its power factor affect the amount of current that the system requires for a given load. The higher the efficiency and the power factor, the lower the current into the system. In some cases these items can make the difference between wire and breaker sizes that will have a significant effect on the installation costs.

The input and load voltages of the system have a large effect on the wiring costs. A 480 V input UPS will draw 43% of the current compared to a 208 V UPS. If possible, all larger systems should be fed from the highest practical utility voltage. The same is true for the load side. In most cases the user does not have good control of the input requirements of the connected loads. However, even if the actual loads are 120/208 V it may make sense to distribute the UPS output at 480 V and then step it down to 120/208 V near the load. This is especially true for higher capacity systems or when there is a long distance between the UPS and the loads.

8.3.4 Cost of Operation Considerations. The following items have the greatest impact on the cost of operation of the system. These items are of general interest to almost all commercial installations.

8.3.4.1 Efficiency. The efficiency of a power system is the relationship between the input power that it draws and the corresponding power that it is able to supply to the load (kilowatt out/kilowatt in). The efficiency of these types of products will vary at different load levels. A value should be obtained for the anticipated load level on the system. The efficiency should reflect the actual operating conditions of a normal system. The conditions under which the efficiency is measured should have all fans, power supplies, etc., operating along with all capacitor bleeder resistors and other power dissipating devices connected. If the efficiency is to accurately reflect the installed conditions, the normal float current into the battery bank should also be taken into account.

The power that is drawn by the system is a function of the load on the system and its efficiency. The poorer the efficiency, the more power will be required to support a given load. This translates into higher operating costs for the less efficient systems. As mentioned earlier, the additional air conditioning costs that result from lower efficiency must also be considered. The less efficient the system is, the more heat it generates. In turn, this heat must be handled by the air conditioning system. This means more energy is required to operate the air conditioning, or possibly an even larger system may be required.

8.3.4.2 Reliability. The overall reliability of the system will impact its total operating cost. A system that is more reliable will typically cost less to maintain and will cause fewer failures in the critical bus. It is sometimes difficult to assess the costs of down time and power-induced failures in the critical loads. These costs vary so much from one installation to another that guidelines are even difficult to create. It is safe to say that unreliable operation can offset efficiency advantages and other performance features of a product. An unreliable system is undesirable no matter what other positive features it may have.

8.3.4.3 Maintenance Costs. All power systems require some preventative maintenance. This will include checking of the electrical connections within the unit as well as the connections between the batteries, cleaning, recalibration, and general diagnostics. If the installation includes wet-cell batteries their specific gravity and voltages should be measured and recorded. These periodic maintenance activities can be covered by the manufacturer's maintenance agreements.

8.3.5 Specifying Engineer's Considerations

8.3.5.1 Operational Specifications. There are a number of operational specifications that need to be considered while specifying a UPS. What follows is a list of those operational specification items that should also be considered.

8.3.5.2 Load Isolation. One of the fundamental functions of a power line conditioner is to prevent its load from being subjected to noise and other disturbances that are present on the input power source. The power line conditioner's ability to isolate the load is usually expressed in decibels (dB) and different values are given for common- and transverse-mode noise. The higher the numerical value of the isolation, the better the load is protected.

8.3.5.3 Input Transient Suppression. This is the amount of transient suppression that the power line conditioner contains to protect itself from large voltage transients on its input. This is usually specified as to the amplitude and duration (which is a measure of the energy contained) of the transient that can be withstood.

8.3.5.4 Overload Capacity and Duration. This is a measure of how much margin is designed into the system. This extra capacity is needed to clear faults and provide additional current for starting various loads. Some products exhibit very poor characteristics, which include high distortion and poor voltage regulation, during periods of overload.

8.3.5.5 Input Voltage Range. This is the range of input voltage that the system can operate over. The system should have full capabilities, including charging the battery over this range of input voltages. The wider the range, the more tolerant the system will be to fluctuation in the input line.

8.3.5.6 Output Voltage Regulation. This specification defines the maximum change in the output voltage that should occur during all transverse modes of operation. It should be specified for all combinations of load changes, input voltages, including the complete loss of input, and during the entire battery discharge. The manufacturers of the equipment on the critical bus should state the acceptable limits for the steady-state regulation for their inputs. These limits should be strictly adhered to.

8.3.5.7 Unbalanced Load Regulation. This specifies the maximum voltage difference between the three output phases that will occur when individual phase loads are different.

8.3.5.8 Output Voltage Distortion. This specification describes the maximum amount of voltage distortion that will be present in the output of the unit when connected to a linear load. A linear or resistive load is one that draws current from its source that is proportional to the voltage waveform. The specification generally defines the total harmonic distortion as well as the maximum value of the largest harmonic that can be present. Most critical loads are not linear loads, so this specification does not reflect the actual distortion when the system is installed and powering the critical bus. In some cases the actual output distortion can be so great that the critical load may not operate. Some power line conditioner and UPS manufacturers specify a value for output distortion with nonlinear loads. However, without defining the exact type of nonlinear load, the specification is meaningless.

One cannot assume that the product that has the lowest distortion specification for a test performed with a linear load will have the lowest distortion in the actual application. This is due to the differences in the output impedance, at the frequency of the distortion, between power conditioners. It is advisable to test the power line conditioner with the intended load if the actual level of voltage distortion is critical. The resulting amount of voltage distortion can be estimated if one knows the amplitude and spectrum of the load's input current distortion and the output impedance of the power line conditioner at those frequencies.

8.3.5.9 Dynamic Response. The output dynamic response of a unit is defined as the deviation that occurs in the output voltage when a load step is applied to the output. Also associated with the deviation is the time that it takes for the output to recover to within normal regulation limits. The specification is an attempt to quantify the disturbance that will occur on the output when a computer or peripheral is started. If the disturbance is too large, the load that is being started or other loads that are being fed from the same critical bus may be adversely affected. The size of the disturbance is usually proportional to the percentage that the load was changed. The recovery time is a measure of how fast the system can respond.

The transient response is usually specified for partial and full load steps. The smaller the deviation and the faster the output voltage recovers to normal the less likely that the loads will be affected. Most computer systems and other critical loads state the maximum transient that they can withstand. The manufacturer's recommendations should be strictly adhered to.

8.3.6 Transfer Characteristics

8.3.6.1 Transfer Time. This is the time that it takes a UPS to transfer the critical load from the output of the inverter to the alternate source or back again. Most of this time is typically used in sensing that there is a need to transfer the load. Once sensed, it takes only a few microseconds to activate the static transfer switch.

8.3.6.2 Automatic Forward and Reverse Transfer. It is important that the UPS system be able to automatically transfer in both directions. Many systems rely on the utility source to supply currents that exceed the capacity of the inverter. In this case the load should automatically be returned to the inverter once the load current has returned to within the inverter's capabilities.

8.3.7 Power Technology Considerations. There are a number of different techniques that are employed by manufacturers today to design the components of the UPS. As new and better semiconductor devices are developed one would expect to continue to see new approaches taken. These technologies are of interest only to the point of what they offer in performance to the user. If a new technology is employed it should have definable advantages to the customer to be worth consideration.

8.3.7.1 Rectifier Technology. For many years the rectifier sections of the UPS have been phase-controlled rectifiers that have employed thyristors to rectify and control the output voltage. These systems are well understood and provide dependable performance. This type of rectifier has several drawbacks that can be improved by proper design. The input current distortion in three-phase units will be around 30% if only a six-pulse rectifier is employed. This high level of input distortion can potentially cause problems with other pieces of equipment that are connected to the same input source. The distortion can be improved by using an input transformer in the UPS that has two secondaries that are phase shifted from each other. When these six phases are rectified by a twelve-pulse rectifier, the resulting distortion should be on the order of 13% or less. This is only achievable if the manufacturer uses thyristors in both sides (positive and negative) of the rectifier bridge.

There are several other types of rectifier sections in use today besides the phase-controlled thyristor units. One type employs pulse width modulation

(PWM) type switching units using transistors. These units can have advantages in the areas of wider input voltage and frequency ranges, higher input power factors, and lower input current distortion.

8.3.7.2 Inverter Technology. Traditionally, most of the inverters in UPS products employ thyristor switching techniques or motor/generators. The thyristor inverters use a number of techniques to control the output voltage and to provide the low-distortion output that is required. These techniques include square-wave, quasi-square-wave, and stepped-wave PWM, etc. Each technique has its benefits in some areas and its limitations in others. The specifications that are typically traded off in these designs are output distortion, transient response, battery discharge voltage range, overload capacity, ability to handle switching loads, and efficiency. Each manufacturer tries to achieve the best combination of the items.

Many older designs typically provide a relatively high-output impedance to the harmonic currents that modern switching loads place on the UPS. This causes the output voltage of the UPS to become distorted. The resulting distortion can have a detrimental effect on the loads that the UPS is supplying. To overcome this and to provide some other performance advantages, the industry is producing inverters that use what are called PWM techniques. The level of performance will depend on how the manufacturer has designed the PWM inverter and its controls.

These methods have been around for quite some time but they were not very successful due to the lack of suitable semiconductor devices. The speed and commutation losses of thyristor circuits do not lend themselves to the high-switching frequencies that are needed to get the full benefit out of a PWM inverter. The industry now has available to it a vast and ever-increasing assortment of both bipolar and MOS-FET transistors that are well suited for the application.

The newer inverter circuits can be smaller and more efficient than the older designs. Also, because of the high-switching frequency of the power circuits, the output of the inverter can be controlled on a subcycle basis. This subcycle control allows the inverter to respond to the current demands of the load that occur at higher frequencies than 60 Hz. A properly designed PWM inverter can respond in a millisecond or so to changes in current requirements of the load. This includes both current distortion and step load level changes. A PWM inverter of this type can provide a less distorted output when connected to a switching load than other methods.

8.4 Equipment and Material Specifications

8.4.1 General Discussion. The purpose of an equipment and/or material specification is to describe technical performance and physical requirements for a piece of equipment or system that is desired by a customer or user. Typically the specification serves as the technical portion of a purchasing contract. The purchase order defines the business terms of the agreement. The most important aspect of the specification is how well the buyer understands the value of what is described. Specifications do not have to be lengthy and complex to be effective. Understanding the values of a well-written procurement specification is the purpose of this subsection.

8.4.2 Using Vendor-Supplied Specifications. The most common method of developing a specification is the use of a manufacturer prepared product specification. Unless a qualified consulting engineer or experienced user is involved, the use of a vendor's printed specification is the only way to give a detailed description of the product desired. However, it must be kept in mind that a particular vendor's prepared specification describes his product in great detail. Other vendors or manufacturers of a similar product will object to this practice since the specification favors another vendor. In some cases, they may refuse to offer a proposal or bid because they feel the buyer has made his choice of product by virtue of the specification used.

The use of vendor supplied specifications does not have to reduce the competitive process. The person responsible for the procurement can promote competitive responses by encouraging other vendors to make a proposal for their similar product. A serious competitor will take the time to respond to another manufacturer's specification pointing out the differences between his product and the specified product.

The use of specifications is essential to the procurement process. The buyer can effectively use vendor-supplied specifications "as supplied" or modified, depending on how well the specification describes the product desired. Careful review of the specification prior to issuance will minimize conflict and maximize the value received.

8.4.3 Creative Specifications. Writing an effective specification for the procurement of a product or service is a difficult task. The writer must first determine exactly what he or she is trying to procure: a specific product for a specific task, or a generic performance criterion. Unless the task is specific or custom in nature, a "performance" type specification will generally provide the best results. By accurately describing the desired performance parameters, more than one vendor or manufacturer can respond with meaningful proposals.

8.4.3.1 Unique or Special Specifications. Writing a specification to cover a unique or special situation should be avoided, if possible. In the majority of cases where special products are deemed necessary to provide a service or solve a problem, a standard product or service actually exists for that purpose. In most cases, it would be beneficial for the custom specification to permit consideration of other approaches to the solution of the requirement. The following are specific examples of problem areas that result from unique specifications.

8.4.3.1.1 Specification Conflicts with Standard Products. The unique specification deviations from standard products can be significant. Major changes may totally discourage vendors from bidding standard products.

Under these circumstances the buyer may have to actually promote the proposal of products. The important task may be to determine if the unique requirements are necessary and that standard products must truly be modified.

8.4.3.1.2 Unknown Performance/Reliability Characteristics. Unique specifications that require extensive modification of standard products may result in reduced performance or reliability in other areas. In the case where a totally new product results, the known "track record" or performance history of a standard product is eliminated. Long-term benefits of standard products may be more important than a unique feature that requires extensive modification.

8.4.3.1.3 Long-term Maintenance Problems. Requiring a unique product by definition means that a "one-of-a-kind" product may result. Typical manufacturers cannot provide their normal degree of engineering or support documentation that accompanies standard products. The most serious consequence is inadequate spare parts and field service once the product is in use. The long-term result may be reduced to unreliable performance later in the product life cycle, or premature replacement of the product in total.

8.4.3.1.4 Electrical Safety Listing Avoidance Problems. One of the most serious ramifications of custom product specifications is the avoidance of product listings (UL, E.T.L., C.S.A., etc.) designed to ensure safe operation. The requirements of these agencies demand extensive testing to ensure compliance with accepted standards. Even the slightest change can sometimes impact the right to list a product and result in not being able to install and operate the product.

8.4.3.1.5 Increased Liability Problems. Use of a unique specification can conceivably increase or involve the purchaser in the liability associated with a failure of a special power system. In the case of liability claims, standard products with a proven track record are the responsibility of the manufacturer if the product was applied or used properly. Development of a unique product may relieve the manufacturer of a portion of his liability if a major problem develops.

8.4.4 "Mixed" Vendor Specifications. When the standard specifications of several vendors are available, there is a strong temptation to select the best features and functions from each vendor and combine these into a single specification. Even though the chosen feature from each source is a standard item for that vendor, the overall specification ends up being very unique. The most typical occurrence of this situation is with functional items, such as operator controls, alarms, status indicators, and other individual components.

Each manufacturer can generally recognize the specific features of his competitors used in the specification. Getting each vendor to address the specification in detail becomes more difficult. Generally too much time and effort are spent on "selling against" items in the specification rather than on user benefits. This approach generally leads to greater confusion on the part of the purchaser, and in turn makes a value selection more difficult.

8.4.5 Generic Specifications for Multiple Vendors. A true generic specification that can be proposed by more than one vendor is possible. A specification that addresses the functions or results desired from a product is typically called a performance specification. This type of specification concentrates on how each system would perform in the critical areas. In the case of a UPS system, as an example, the main performance issues would be the following:
  (1) Rating (kVA/kW)
  (2) Dynamic response (voltage regulation)
  (3) Overload capability
  (4) Efficiency
  (5) Failure-mode performance

In a "performance" specification the hardware items allow vendors to present the ways in which their product satisfies the specification. The support items or "software" (test procedures, quality assurance, maintenance agreements, etc.) can be tailored to the specific project. The items determine the degree of support required on a project-by-project basis.

8.5 Verification Testing. This section generally applies to large systems where the effort and expense of verification testing is justified. For these systems, there must be some method devised to determine that the product being procured does in fact meet the specifications for which it was purchased. This function is usually performed through acceptance testing performance at the manufacturer's facility before shipment and on site after installation. The manufacturer can supply what he or she considers to be a valid test of the product's performance which then can be modified to cover those items of particular importance to the specific installation.

The factory acceptance should verify that the power conditioner meets all of its significant specifications in the environment of the manufacturer's test facility. The on-site acceptance should verify that the system has not been degraded by the transportation and installation at the new site. It should further verify that the system will function properly in its new environment with the actual load that it was intended to support. This phase of testing is very important because it determines if all of the effort that went into the specifying and earlier testing has actually resulted in a system that will perform the desired function.

The following describes some of the most important tests that should be performed in general terms. The unit under test is assumed to be a UPS. Delete or modify those tests that do not apply to the type of equipment being tested.

After final assembly and quality control (including normal factory tests) are completed, the manufacturer should perform the following UPS tests witnessed by the customer's representative.

8.5.1 Visual Inspection. A qualified individual can gain insight by simply looking at the components used in the power conditioner and the methods and workmanship of assembly. The trained individual should inspect to see that the cabinets are of adequate strength to withstand the stresses of transportation, installation, and seismic activity. Components should be high quality and properly mounted to assure mechanical security and adequate heat transfer. The wiring should be of the proper rating, properly terminated, and secured to prevent damage. Bus bars should be properly mounted and braced to resist movement during fault conditions. These areas and others can tell a significant amount about the long-term reliability of the product.

8.5.2 Load Tests. This test should be performed to verify that the UPS is correctly connected and all functions operate properly. The test should include adding blocks of load in 25% increments to full load at specified power factor. Observe and record the output voltage amplitude, waveform, steady-state regulation, surges, and frequency. Check the operation of all controls, meters, and indicators.

8.5.3 Transfer Test. This test should be performed to verify that the UPS will transfer from the inverter to the alternate source and back without generating disturbances on the load bus beyond specified limits. At no load and at full load manually transfer the load to bypass source and then back to the UPS. Observe and record the same parameters as in the load test above. In addition, the transfer time in each direction should be determined and recorded.

8.5.4 Synchronization Test. This test should be performed to verify that the UPS is able to synchronize to alternate sources within the specified limits. The frequency of the alternate source should be varied outside of acceptable limits. An attempt to manually transfer to this source should be made. The alternate source should then be returned to nominal frequency and, after the specified synchronization time, a manual transfer should be attempted. The results of these tests, including surges on the output, should be recorded.

8.5.5 AC Input Failure and Return Test. This test should be performed to verify battery operation. Perform this test by interrupting and restoring the ac power source to the UPS. The steady-state voltages, transient voltages, and frequencies should be recorded. The system should be allowed to operate from the battery, at rated load, to determine performance and specified battery time.

8.5.6 Efficiency Test. This test should be performed to verify that the UPS is operating at the specified level of efficiency. The ac-to-ac efficiency of the UPS shall be measured and recorded at full and partial loads. This is done by measuring the real power input and output and dividing the two figures.

8.5.7 Load Performance Test. This test should be performed to verify that the UPS has specified capability at the extreme operating conditions. Tests should be performed at full load and rated power factor with the output voltage set to its maximum rated level and at the lowest specified dc bus voltage. Record the output voltage and frequency.

8.5.8 Load Imbalance Test. This test should be performed to verify that the UPS is capable of supplying unbalanced loads per specification. For three-phase systems, the phase-to-phase, and the phase-to-neutral voltages and phase displacements should be measured and recorded with a balanced full load on the UPS output. The maximum specified load imbalance should be applied and the same parameters should be measured and recorded.

8.5.9 Overload Capability Test. This test should be performed to verify that the UPS is capable of supplying specified overloads. The maximum specified overloads (current and time) should be applied to the UPS and its output voltages and current should be measured and recorded.

8.5.10 Harmonic Component Test. This test should be performed to assure that the UPS does not generate harmonics in excess of specification. The harmonic content of the input current and the output voltage should be measured and recorded at full rated load.

8.6 Equipment Maintenance

8.6.1 Preventative Maintenance. It is generally accepted that equipment with moving parts requires periodic maintenance in order to assure reliable operation. Such items as cleaning, lubrication, and adjustments for wear are common in the upkeep of mechanical equipment. What may not be as obvious is that power electronic equipment requires periodic maintenance as well. A proper schedule of periodic inspections will enhance the equipment's reliable operation.

The following list outlines some of the operations that are performed during preventive maintenance of power conditioning equipment:
  (1) Check security of all electrical connections (including batteries).
  (2) Clean units and batteries and replace air filters.
  (3) Check battery cell voltages and specific gravity (wet cells).
  (4) Lubricate components as required.
  (5) Visually check power connections and components for signs of
      overheating, swelling, leaking, etc.
  (6) Perform calibration of meters, alarm levels, etc.
  (7) Functionally check the operation of all components.
  (8) Perform system performance checks.

The list above is for illustration only; the manufacturer's recommendations should be followed strictly. By performing this type of maintenance on a scheduled basis, it is possible to find and remedy potential problems before the system's operation is affected.

8.6.2 Wear and Aging of Components. We have come to expect that mechanical components wear during operation. This wear can usually be seen or measured. Some electrical components "wear" during operation as well, but it is sometimes more difficult to detect.

Rotary or M-G products experience wear in their bearings and, in some cases, brushes. Fan motors also experience bearing wear. Breakers, switches, and contactors experience wear in the mechanisms as well as the electrical contacts. Many components, such as motors, transformers, and capacitors, experience degrading of internal insulation over their life.

The rate of this degrading with time is a function of the design of the component and the level of stress to which it is subjected. A given component may have a much longer operational life in a conservatively designed product than it would in a design where its stress level is higher. The design stress level of a component is related to how close the component is operating to the manufacturer's maximum specifications. Typical parameters involved include peak voltage, rms current, and temperature and power limits. In most cases, the designed stress level interacts with the operational environment to determine the ultimate life of the component. High-temperature environments tend to shorten the life of nearly all components. The life of some components, such as electrolytic capacitors and batteries, are greatly affected by operation at elevated temperatures.

8.6.3 Restoring System Operation After Failure. There will be failures even in a well-maintained system. When failures occur, it is important to take the proper steps to restore the system operation as soon as possible. The following lists the general order of events that should occur when there has been a failure:

(1) Determine what has failed and why it failed.
(2) Restore power to load through the use of maintenance bypass switchgear or other means.
(3) Replace or repair the failed component or assembly.
(4) Restart system and perform operational checks.
(5) Place system back in service.

If the critical load has lost power, the first step is to restore power. This is often performed through use of bypass switchgear that will connect the utility power directly to the load. It is generally advisable to close a manual bypass switch even if the load is being supplied through a static switch or other automatic switch.

Clearly the next step is to determine what has failed in the system. Modern power-conditioning systems provide alarm annunciation and some provide effective diagnostics to help identify the source of the problem. The ease of determining what has failed and the actual repair of the system varies with its design. It is typically easier to isolate the problem and to replace complete assemblies as opposed to individual components. System designs that have made good use of modular repair concepts will generally be easier and faster to put back in service.

The second part of this step is to determine why the failure has occurred. There is normally a cause for each failure, and it needs to be determined and dealt with to avoid recurrences of the same failure. This can be difficult because the cause is often transient and no longer present. The source of the problem could be internal to the equipment, in the utility feed, the building power distribution, or the load itself. It is often not possible to devote the time necessary to determine the cause because of the need to restore the system to operation. In that case, steps should be taken after the system is in service to determine and eliminate the source of the failure.

Once the failed part or assembly is identified and repaired or replaced, it is advisable to perform sufficient operational tests to assure that all areas of the system are now functioning properly. Other components may have been damaged and need to be repaired. Once the system is fully checked out it can be placed back in service. Accurate records of the failure and all associated data should be kept to aid in any future correlation of this failure with others. The actual cause of the failure may not be determined until the data from this failure is compared to other failure data and operational records.

8.7 Bibliography

[B1] Federal Information Processing Standards Publication 94: *Guideline on Electrical Power for ADP Installations*, Sept. 21, 1983.

[B2] IEEE Std 446-1987, IEEE Recommended Practice for Emergency and Standby Power Systems for Industrial and Commercial Applications (IEEE Orange Book).

[B3] IEEE Std 449-1990, IEEE Standard for Ferroresonant Voltage Regulators (ANSI).

Chapter 9
Recommended Design/Installation Practices

9.1 General Discussion. This chapter deals with good engineering principles that relate to performance requirements of modern electronic equipment. Recommended practices may appear to restrict design, installation, or service efforts. These restrictions generally are necessary to obtain the desired performance levels of sensitive electronic loads within the confines of the applicable codes, standards, and regulations. Sensitive load installations, large and small, may be affected by multiple codes, standards, and regulations. Therefore, associated standards and regulations, beyond the applicable electrical code; e.g., the National Electrical Code (NEC) (ANSI/NFPA 70-1993 [2][25]); should be used by the prudent designer, installer, and service person.

9.1.1 Safety. Electrical safety is the overriding concern of all electrical design work. All other aspects of the job should be of secondary importance to the issue of safety. Safety is basically governed by the electrical codes and standards as adopted by government agencies, commercial entities, and good judgment on the part of the designer.

In cases where more than one performance and/or safety design alternative exist, as described by several codes and standards that may apply to a given situation, preference should be given to those that have been adopted as mandatory by the governmental authority having jurisdiction over the job. Next in priority should be the applicable nongovernmental electrical safety requirements (e.g., internal or corporate standards and supplier instructions). These documents are followed in priority by those recommended electrical safety practices that remain and are governed by sound engineering judgment and experience.

Disputes arising from the interpretation of governmentally imposed electrical safety requirements, codes, or standards should always be decided by appeal to the governmental authority enforcing the subject requirement, code, or standard. These matters should not be resolved by ad hoc consensus, unqualified persons, or by persons with a potential conflict of interest.

In general, equipment that cannot be made to operate in a satisfactory manner without violating applicable electrical safety requirements is not suitable for use in normal applications. This inability is considered to be a design flaw of the subject equipment. As such, the equipment should be properly modified by its original equipment manufacturer (OEM), or by OEM authorized field service or engineering personnel so that it will work in a safe manner.

---

[25] The numbers in brackets correspond to those of the references in 9.21; the numbers in brackets, when preceded by the letter B, correspond to those in the bibliography in 9.22.

It should not be placed into service if unsafely wired or if otherwise unsafely installed.

The exclusive use of electrical and electronic equipment that is covered by a product safety test or listing is generally the first line of defense against electrical safety problems. With very few exceptions, the use of listed equipment is also required by applicable electrical codes, such as the NEC [2]). Listed equipment is required by the listing agency to be compatible with applicable electrical codes. Listed equipment shall only be utilized for the specific applications for which it is listed.

9.1.2 Performance. The performance of electronic equipment is often closely tied to the method of equipment installation. Of particular interest in this respect are the requirements for external ac power and grounding for the connected equipment. Equipment and ac system grounding incompatibilities are the most common electrical wiring problems encountered. Great variation exists among OEM requirements for these installation parameters. In certain instances, OEM requirements conflict with the applicable safety requirements for the installation. When this situation occurs, safety requirements must take precedence over the equipment's performance.

9.1.3 Three-Phase Versus Single-Phase Systems and Loads. Some power conditioning and electronic load equipment are operable *only* from a three-phase power source, while other equipment may require *only* single-phase. Often single-phase equipment can be operated directly from a single-phase component of a three-phase system. However, these alternatives should be carefully determined before setting forth an electrical system design.

When evaluating the ability of load equipment to operate on either a single-phase or three-phase ac system, it is often necessary to determine if the equipment has internal taps or other adjustments that may affect a voltage-level change, or if the equipment simply has enough range to accommodate the two nominal supply voltages without internal adjustments. Much modern equipment falls into the latter category.

When evaluating the choice between three-phase and single-phase systems, consideration should always be given to the fact that three-phase systems may generally support larger loads with greater efficiency than single-phase systems, and that single-phase power may be derived from three-phase systems. The derivation of three-phase power from a single-phase system is not practical. Certain methods of converting a single-phase circuit to supply three-phase loads (e.g., capacitor phase-shifters) are inappropriate for electronic loads and can damage these loads (IEEE Std 141-1986 [11]).

9.1.4 Selection of System Voltages. The selection of the ac supply system voltage typically begins at the service entrance (SE) of the facility. Choosing a service and interior distribution voltage that is higher than the actual utilization-equipment level of 208 or 120 Vac is generally desirable. Higher voltages (e.g., 480 Vac) are less susceptible to on-premises generated disturbances, and are generally more stable than lower voltage systems when derived from the same electrical utility feeder.

The higher voltage supplied system is also generally provided with more fault-current capability as a result of fewer (and lower) series impedances between the electric utility's supply and the served loads. Higher voltage systems also benefit the user in terms of lower A/kVA and higher kVA/$. Better overall efficiencies of the higher voltage wiring systems, due to lessened IZ drops and other electrical loss-related considerations, also make the higher voltage distribution systems desirable.

9.1.5 AC System Waveforms. Modern electronic load equipment tend to contain nonlinear loads, and require high peak-currents with attendant large crest-factors from their ac supply circuits. These loads are capable of creating significant harmonic voltage-waveform distortion and, when combined, may affect the entire premises' electrical distribution system. In some cases, there may be a considerable local effect on the electric utility system.

In certain situations electronic load systems may not be the actual cause of voltage-waveform distortion problems, but may be the victim. This situation generally occurs when large nonlinear load equipment distorts the voltage waveform of a building supply circuit and all other building loads are therefore affected. IEEE Std 519-1992 [15] is the recommended reference for these issues.

9.2 Computer Room Wiring and Grounding. Creation of a computer room that meets the requirements of NEC [2], Article 645, permits the designer to utilize flexible wiring methods within the room that would otherwise not be permitted. Related design information is also presented in ANSI/NFPA 75-1992 [3]. It is recommended practice that a computer room, per the NEC, Article 645, and ANSI/NFPA 75-1992 descriptions, be created and maintained where large sensitive electronic systems or automatic data processing systems are to be installed.

9.2.1 ANSI/NFPA 75-1992 [3]. This document provides specific definitions of interconnecting cables and other items used in conjunction with the NEC [2]. It also cross-references the NEC, ANSI/NFPA 780-1992 [4], and numerous other important NFPA references. ANSI/NFPA 75-1992 does not apply to areas other than designated computer rooms and their directly related support areas (e.g., media storage areas).

9.2.2 UL 1950-1989 [20]. This standard has provisions for listing power conditioning, distribution, and control equipment that are—
(1) Connected by branch circuits (not feeders) under 600 Vac rating,
(2) Not installed as a part of the premise mechanical or electrical systems, or
(3) Installed only as a UL 1950-1989 listed part of a listed electronic computer/data processing system that is comprised of a single or multiple vendor-provided set of electrical or electronic load units.

UL 1950-1989 also contains the listing requirements for all interconnecting cables for listed units of the electronic computer/data processing system. Cord assemblies and interconnecting cables listed to this standard are specifically stated to be suitable for installation within the space under a cellular raised floor, with or without that space being used for heating, ventilation, air conditioning, and process cooling airflow (see 1.5 of UL 1950-1989).

9.3 Dedicated and Shared Circuits. When supporting simple loads, it is common practice to have loads of unlimited variety share both feeder and branch circuit wiring. This approach is based on economics as the principal driving force, and normally there is little fear of load incompatibility on shared circuits with simple loads. However, such is not the case with electronic load equipment where unwanted interactions can cause performance and reliability problems. Sharing common line, neutral, or grounding conductor paths by multiple electronic loads may produce unwanted interactions.

9.3.1 Typical Forms of Unwanted Interaction on Shared Circuits. One form of unwanted interaction is where one load faults and causes operation of the ac supply circuit overcurrent device, thus shutting down all loads on the shared circuit. Other forms of unwanted interaction may be complex and difficult to diagnose. These complexities often include position sensitivity in which loads may be moved about on a shared circuit to find compatible and incompatible positions. Unwanted transient voltages and currents can intermittently appear at tapping points on multi-outlet assemblies in response to $L\,di/dt$ effects. Such events are often initiated by switching operations on the ac supply system and by the effects of lightning currents.

9.3.2 Dedicated Load Circuits. Recommended practice of wiring system design for sensitive electronic load equipment is to separate dissimilar classes of loads by placing them on dedicated feeders and branch circuits, and not combine them except where they are known to be compatible. Support equipment (e.g., heating, ventilation, air conditioning, and process cooling equipment) should be powered via a separate feeder and panelboard-branch circuit system (Fig 9-1).

9.4 Feeders. Feeders are not used to directly connect load equipment, but are used to route bulk ac power between a source and a distribution or transformation point. Feeders are also used between distribution points, such as switchboards or panelboards and disconnects. Feeders may be quite large and lengthy, and are generally capable of delivering large fault currents under short-circuit conditions involving ground. Therefore, careful attention should be paid to the control of voltage drops in these circuits and for proper grounding.

9.4.1 Voltage Drops in Feeders. Voltage drops in feeders that serve sensitive electronic loads (or associated power conditioning equipment) should be no more than 2% under the actual conditions of connected load.

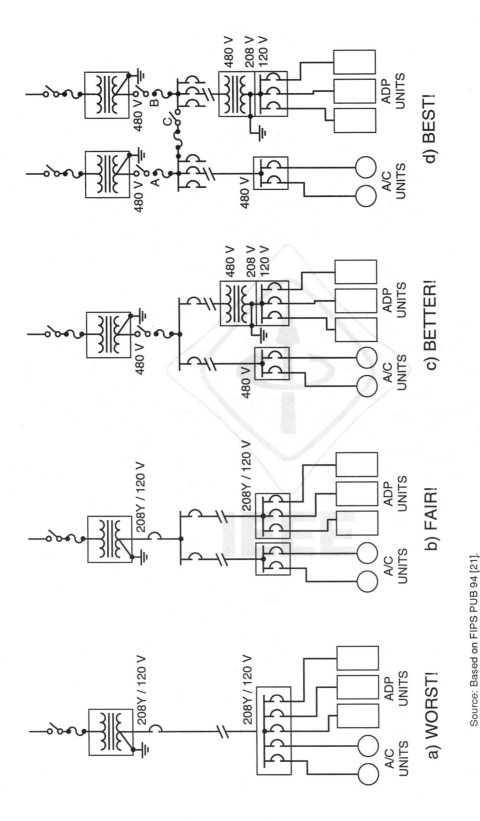

Fig 9-1
Recommended Separation of Sensitive Equipment Power Distribution From Support Equipment Power Distribution

Source: Based on FIPS PUB 94 [21].

9.4.2 Current Ratings for Feeders. Typical electronic load equipment is continuous running, nonlinear, and automatic voltage regulating. Therefore, feeders directly serving these loads should be rated for the higher associated phase currents (due to inverse voltage-current load characteristics) and neutral currents (due to nonlinear characteristics).

9.4.3 Busway Configurations. A fully enclosed, dedicated busway (without taps) is recommended practice. This design avoids problems that often result from multiple loads being connected to the same feeder along its length and that may interact with one another via the commonly shared wiring impedances. If a nondedicated busway is used (with taps), then a separately derived source should also be configured at each tap that serves sensitive equipment.

9.5 Branch Circuits. Branch circuits are always the final interface between the premises' wiring system and the load equipment.

9.5.1 Voltage Drop on Branch Circuits. The voltage drop on branch circuits serving sensitive loads should not exceed 1% of the line voltage.

9.5.2 Shielding of Branch Circuits. Branch circuits are recommended to be installed in continuously metallically shielded conduits and raceways.

9.6 Avoiding Single-Phase Input Conditions on Three-Phase Load Equipment. Most three-phase electronic load equipment cannot tolerate the application of single-phase power to its input. The resulting downtime and equipment damage can be extensive. Because fuses and circuit breakers generally cannot prevent all types of single-phasing conditions, electronic phase-failure or voltage-unbalance relays are recommended in addition to fuses or circuit breakers.

Proper application of fuses and circuit breakers requires that the overcurrent device time-current curves be coordinated and matched to the load characteristics. Where significant inrushes are expected, time-delay devices are recommended. Where harmonic currents or distorted wave shapes are encountered, true rms sensing devices such as fuses or solid-state true rms sensing circuit breakers are recommended.

9.7 Harmonic Current Control on the AC Supply Wiring System. Refer to IEEE Std 519-1992 [15] for a general discussion of harmonic currents. Recommended practice is that all power distribution systems intended for use with sensitive electronic loads be designed per IEEE Std 519-1992 guidelines.

9.8 Power Factor Improvement. Low power factors produced by general utilization and electronic load equipment at the site may need to be corrected. Power factor correction should be applied directly at or close to the facility service-entrance due to economics and ease of design. This recommendation applies particularly to sites having large proportions of nonlinear loads that produce considerable harmonic current. Recommended practice is to follow the guidelines given in IEEE Std 399-1990 [13] and IEEE Std 519-1992 [15].

9.9 Specialized AC Source Transfer Switching. This subject is thoroughly addressed in IEEE Std 446-1987 [14]. Recommended practice is that all emergency and standby systems intended for use with sensitive electronic loads be designed per this standard.

9.10 Grounding, General. All metallic objects on the premises that enclose electrical conductors or that are likely to be energized by electrical currents (e.g., circuit faults, electrostatic discharge, and lightning) should be effectively grounded for reasons of personnel safety, fire hazard reduction, protection of the equipment itself, and performance. Solidly grounding these metallic objects will facilitate overcurrent device operation (ground faults), and permit return currents from EMI filters and surge suppressors, connected line-to-ground or line-to-chassis, to flow in proper fashion.

9.10.1 Annotating Mechanical and Electrical Drawings. It is recommended that all grounding and bonding connections for metal piping systems be noted on the appropriate mechanical and electrical drawings. Drawings of existing buildings should also be reviewed for grounded and bonded items that are not properly noted.

9.10.2 Solidly Grounded AC Supply Systems. Recommended practice is to always utilize solidly grounded ac supply systems that support sensitive electronic load equipment. All metal equipment parts such as enclosures, raceways/conduits, and equipment grounding conductors (EGCs) and all earth grounding electrodes should also be solidly joined together into a continuous electrically conductive system.

All metallic systems should be solidly interconnected to the electrical system as provided by the service entrance and for each grounded separately derived ac system (SDS) that is installed. Specific metallic systems included in this requirement are the main and interior cold-water piping system, the structural building steel system, and any other earth grounding electrodes that may be present on the premises.

9.10.3 Working With Improperly Grounded Equipment. Recommended practice is that all grounding design and installation conform to all applicable codes and standards. Refer to (NEC) [2] and IEEE Std 142-1982 [12] for safety grounding.

9.10.4 Grounding of Building Structural Steel. Where it is accessible, all structural steel should be electrically grounded and bonded into a single, electrically conductive mass. Such grounding and bonding may be by structural means, such as welding, bolting, riveting, or by grounding and bonding jumpers. Earthing of the structural building steel system is also recommended.

The structural building steel system should be bonded to the grounded conductor of the incoming ac supply system at the service entrance as well as to the equipment grounding conductor system and the main (metallic) cold-water piping system. Structural building steel should be grounded by one or more of the following means:
  (1) By a made earth-grounding electrode system and grounding or bonding jumpers.
  (2) By direct burial or concrete-encased structural building steel electrode in foundation footings.
  (3) By buried ring-ground electrode system and grounding or bonding jumpers.

9.10.5 Buried Ring-Ground Electrode System. It is recommended that the sensitive electronic load facility be provided with a buried ring-ground. This ring-ground should be frequently bonded to the structural building steel system and to any electrical and piping systems that cross it. Such bonding should always occur at the most immediate point to the intersection between the ring-ground and the item being bonded, to limit the inductance of the connection. The buried ring-ground should also be bonded to lightning down-conductors and to any lightning or other earth grounding electrodes that may be present on the premises (ANSI/NFPA 780-1992 [4]).

9.10.6 Bonding Across Building's Interior/Exterior Line of Demarcation and a Buried Ring-Ground. A building's six sides (comprised of walls, foundation, and rooftop) generally make up the line of demarcation for grounding and bonding of all metallic items entering/leaving the building. It is often convenient to make low-inductance connections at ground level, from the buried ring-ground to metallic items entering/leaving the building.

9.10.7 Grounding Mechanical Equipment in Electronic Areas. All mechanical equipment in the electronic equipment areas should be effectively grounded for electrical safety (according to the NEC [2]), for lightning protection (according to ANSI/NFPA 780-1992 [4]), and for noise-current control. Such equipment (including the structural framing, chassis, piping, ducting, electrical conduit/raceways) should be grounded or bonded to local structural building steel using direct or high-frequency (HF) grounding or bonding means.

When located in the same area as the sensitive electronic load equipment, mechanical equipment should be bonded to the same HF ground reference structure as the sensitive electronic load equipment. Heating, ventilation, air conditioning, and process cooling equipment, related metal piping, and electrical conduits are recommended to be bonded to the local signal reference structure (SRS) where the electronic load equipment is installed.

9.10.8 Associated Electrical Conduits/Raceways and Enclosures. For the purpose of shielding, all metallic conduits and raceways in areas containing sensitive electronic equipment should be bonded to form an electrically continuous conductor, whether or not a separate equipment grounding conductor is provided with the circuit.

9.10.9 Bonding Across Expansion Joints in Electrical Conduits. Bonding jumpers should be placed across expansion joints under all conditions.

9.10.10 Grounding of AC Services and Systems. AC services and systems are grounded for both electrical safety and fire safety reasons as well as for performance reasons relating to the control of common-mode noise and lightning current. From a performance standpoint, when ac systems are supporting sensitive electronic equipment, solid grounding is recommended practice on ac systems to ensure the existence of a conductive path for the return current of LC filters and surge suppressors connected line-to-ground or line-to-chassis. It is also recommended to design for the lowest reasonable impedance between the load equipment containing an LC filter or TVSS and the associated ac supply source.

9.10.11 Grounding of Separately Derived AC Sources. If a typical dry-type transformer or similar ac source (e.g., inverter winding, alternator) is to be installed, this SDS (transformer secondary) should be grounded to the equipment grounding conductor and to the nearest effectively grounded structural steel. If no effectively grounded building steel is available, then the ac system should be connected to the service entrance grounding point.

9.10.11.1 Solidly Interconnected AC System Grounding. Solidly interconnected forms of ac systems grounding should be used when two ac systems supply a common output via a transfer switch arrangement. This subject is discussed in detail in IEEE Std 446-1987 [14].

9.10.11.2 Recommended Interface of the Solidly Interconnected AC System to Sensitive Electronic Equipment. The recommended interface of the solidly interconnected ac system arrangement (with served sensitive electronic load equipment) is a locally installed and referenced isolation transformer (IT) (Fig 9-2). The IT is the recommended form of interface when two ac systems are both not referenced to the same HF ground reference as the served sensitive electronic load equipment.

9.10.12 Isolated/Insulated Grounding (IG) Method. The IG form of grounding is only intended to be used as a possible means of obtaining common-mode electrical-noise reduction on the circuit in which it is used. It has no other purpose and its effects are variable. Results from the use of the IG method range from no observable effects, the desired effects, or worse noise conditions than when standard solid grounding (SG) forms are used on the sensitive equipment circuit.

IEEE
Std 1100-1992   CHAPTER 9

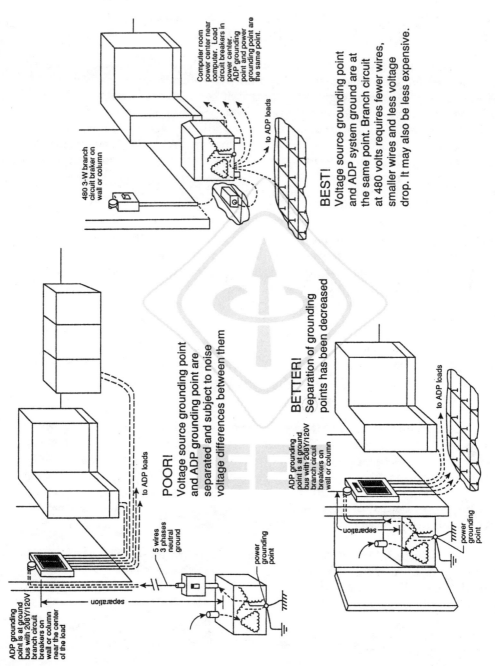

Fig 9-2
Best Design Locates Shielded Isolation Transformer as Close to Sensitive Loads as Possible

Source: Based on FIPS PUB 94 [21].

9.10.12.1 Wiring Means Used With IG Method. The IG method is only directly applicable to metal-enclosed wiring means; it has no special purpose with nonmetallic wiring systems. Nonmetallic wiring systems are inherently constructed as if they are IG, since no metal conduit/raceway is involved in the wiring path to be interconnected to the circuit's EGC.

All nonmetallic wiring systems (e.g., those constructed with type "NM" (nonmetallic) wiring or plastic conduits/raceways) and nonmetallic device mounting boxes constitute an IG design when combined with the SG form of receptacle at the branch circuit's end. An exception is when the circuit outlet is locally bonded to a metallic path that is also part of the site grounding structure.

9.10.12.2 Special Forms of Earth Grounding Electrodes. The use of any separate, isolated, insulated, dedicated, clean, quiet, signal, computer, electronic, or other such improper form of earth grounding electrodes for use as a point of connection of the IG EGC is not recommended. These improper IG grounding schemes do not meet code requirements for effective grounding. The generally perceived need for an *isolated* earth grounding electrode scheme in relation to the IG method is not based on good engineering practice. Isolated earth grounding electrode designs have no means for limiting the potential developed across the intervening impedance in the commonly shared grounding medium (e.g., earth) when a current is caused to flow through it. As a result, lightning may create conditions of several thousands to tens of thousands of volts between two (or more) such earth grounding electrodes. AC system ground faults may create similar problems in relation to the ac system's nominal line-to-ground voltage and the fault-current magnitude FIPS PUB 94 [21].

9.10.12.3 Externally Routed IG Conductors. All IG conductors should be routed within the same metallic conduit/raceway as their associated circuit conductors for the entire length of the involved circuit. Terminations of the IG conductors similarly should remain within the associated equipment enclosure. Failure to adhere to this recommendation will significantly increase the IG EGC effective impedance during fault conditions.

9.10.12.4 Special Metallic Raceway Grounding and Bonding Requirements With IG Circuits. Some conduit/raceway and cable sheath materials (e.g., flexible metal conduit >6 ft long) are not suitable for equipment grounding paths without being bonded across by an EGC.

9.10.12.5 IG for Receptacle Style Branch Circuit End Terminations on Metallic Wiring Systems. Branch circuit outlet receptacles on metallic wiring systems may be configured as IG forms of circuits. This requires the use of IG receptacles in which the EGC pin is insulated or isolated from the device's metal mounting yoke.

The designer is allowed to choose the point between the ac source and its related IG receptacle at which the receptacle EGC pin and the metal conduit/raceway or equipment enclosure system are made common (Fig 9-3). Such a connection must conform to all applicable codes. In general, the choices for grounding a receptacle, its associated branch circuit and feeder wiring (that are not bonded to the metal frame/enclosure of the equipment) are at the directly associated panelboard, switchboard, or the source-ac-system for the IG circuit. Thus, an IG arrangement may be continued from the receptacle upstream to a point no further than the ac source for that receptacle.

9.10.12.6 Recommended Design Procedure for IG Circuits. The following paragraphs set forth a recommended practice for the design of typically encountered forms of IG branch circuits.

9.10.12.6.1 Color-Coding IG Circuits. The IG EGCs should have green-colored insulation with a longitudinal yellow stripe. Black insulated IG conductors (typically larger than #6 AWG) should be color-coded by adding a combination of green and yellow tapes, applied next to each other, at both ends of the conductor. Color-coding should also be added at all visible or accessible locations along the IG conductor length.

Direct-connected (hardwired) circuits employing the IG conductors should have their metal conduit/raceway or cable sheath prominently and permanently identified as being IG. This identification should be made by labeling with an orange triangle symbol or by finishing both ends of the circuit with an orange color. IG-style receptacles on branch circuit outlets should be labeled by an orange triangle symbol permanently affixed to the visible face of the IG receptacle or by being orange-colored.

9.10.12.6.2 Terminating Mixed IG And SG Circuits Within the Equipment. Switchboards or panelboards may require both IG and SG terminations. The same bus-bar logically cannot be used for both under all conditions, e.g., where the IG circuit is continued upstream through a panelboard. However, if the panelboard is actually the termination point for the IG conductor, then it is possible to use the same grounding bus-bar for both the SG and IG conductors.

A common situation is where only the branch circuits are IG and SG styles, and they are terminated within the panelboard containing the overcurrent protection for these branch circuits. In this case, common or separate IG and SG grounding bus-bars may be provided within the same panelboard. Separate IG and SG grounding bus-bars facilitate the convenient current measurement of the total IG current to the panelboard metal enclosure via a (low-inductance) grounding jumper between the bus-bars.

9.10.13 High-Frequency (HF) Ground Referencing Systems. An SRS should be employed as the basic means of achieving a good HF ground reference for all equipment within a contiguous area. The SRS typically can be constructed in the form of a signal reference grid (SRG). The use of a signal reference plane (SRP) is recommended for some applications where the subject system operates at a higher frequency than the typical SRG design cutoff frequency.

RECOMMENDED DESIGN/INSTALLATION PRACTICES

IEEE
Std 1100-1992

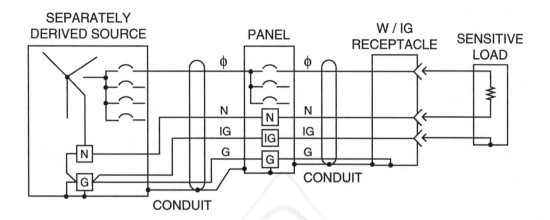

Fig 9-3(a)
Isolated Grounding Conductor Pass-Through Distribution Panel

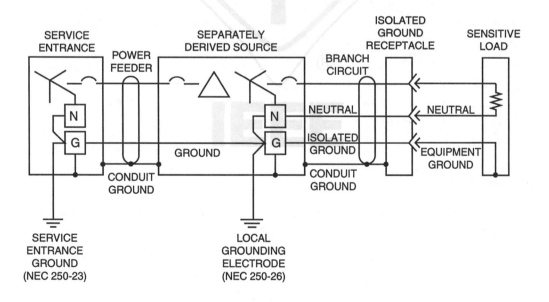

Fig 9-3(b)
Isolated Grounding Conductor Wiring Method
With Separately Derived Source

Hybrid forms of SRS employing mixtures of SRG and SRP for varied construction and HF performance are also useful. They are used where the benefits of each type are needed for the collective support of sensitive electronic load equipment that is susceptible to common-mode noise current related problems.

HF grounding for data signaling cables between (noncontiguous) areas can typically be accomplished via the use of conduit or electrically continuous, solid-bottom, metal cable tray (MCT) or similar forms of construction. These items should be used with supplementary grounding paths (e.g., bonding to structural building steel or steel structural subfloor decking).

Recommended practice for HF referencing of sensitive electronic load equipment does not involve the earth or any earth grounding electrode system except for safety and lightning related surge control. Earth and earth-related paths are not a desired part of the signal process path system. HF grounding principles are discussed in Chapter 4.

9.10.13.1 SRS in a Defined Area. The use of properly bonded screening or solid metal sheet is an ideal form of SRS, but is often expensive. A more practical form is the SRG, which is recommended practice for larger sensitive electronic equipment installations. The SRG is typically configured via either a raised-flooring understructure (if a raised flooring system is used) or a grid of metallic strips or wire.

9.10.13.1.1 Raised Flooring Understructure SRG. A simple and effective SRG is the bolted-stringer understructure of the raised flooring system (Fig 9-4). This SRG can be effective over a broad frequency range. The flooring stringers are typically on 2 ft centers. The inductive reactance of the stringers is usually low. The closer the grid spacing, the lower the inductive reactance across the entire SRG. Bolts connecting the stringers at each pedestal should be maintained tight and corrosion-free. Typical joint resistances of 500 $\mu\Omega$ can be obtained by proper torquing of these bolts. Raised flooring with no stringers, lay-in stringers, or snap-in stringers are not recommended as an SRG. All connections for this form of SRG should be made so no conduction path is required through the threaded pedestal connections because of their unstable contact resistances. A significant advantage of this type of SRG is that the length of bonding straps to protected equipment and associated metallic objects can be very short.

9.10.13.1.2 Flat Strip SRG. Several manufacturers supply an SRG, based on flat copper strips welded or brazed at the crossovers (Fig 9-5). These SRGs can be prefabricated or field assembled and, when crossovers are welded or brazed, generally do not require routine maintenance. The grid lays directly on the subfloor that supports the raised flooring. Power and data cables lay on the grid. The advantage of this geometry is that the coupling of radiated energy into the cables is minimum when they are very close to the ground plane [B3]. The higher capacitance between the cables and the SRG increases the protected circuit's noise immunity to electric fields. Minimum

RECOMMENDED DESIGN/INSTALLATION PRACTICES

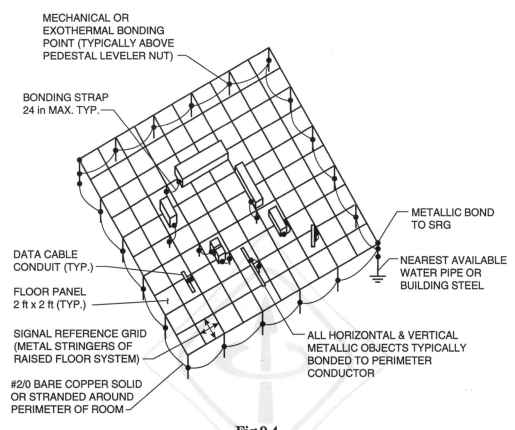

**Fig 9-4
Raised Access Flooring Substructure as Signal Reference Grid**

spacing between the cables and the SRG also reduces susceptibility to magnetic fields. A disadvantage of this form of SRG is the requirement for longer bonding straps compared to the raised-floor based SRG. Two bonding straps (of different lengths) to each piece of equipment reduces the bonding strap impedance substantially.

9.10.13.2 Grounding the SRG. Safety requirements dictate that the SRG be connected to an appropriate earth ground. The ground connection has no relationship to improved system performance. This is an important point that has caused a great deal of confusion and has led to poor design practices.

9.10.13.2.1 Multipoint Grounding of the SRG. The recommended practice for most facilities is multipoint grounding of the SRG. Multipoint grounding requires that all metallic objects crossing the SRG are bonded to it. This recommendation includes all building steel and other conducting paths within 6 ft of the SRG. Concrete-encased steel is considered to be nonaccessible, so a separation of 6 ft is required. In new construction, concrete-encased

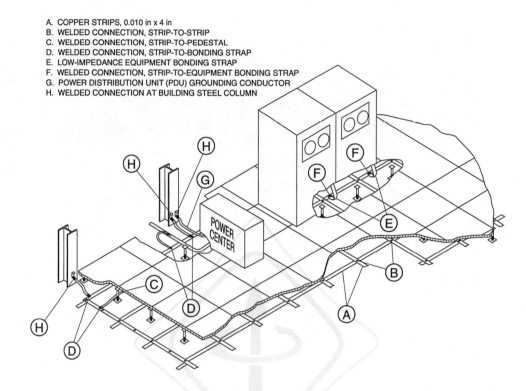

**Fig 9-5
Signal Reference Grid Fabricated From Copper Strips**

steel should be bonded to the SRG. This form of grounding requires minimum maintenance.

9.10.13.2.2 Single-Point Grounding of the SRG. Single-point grounding is not a recommended practice. Although the single-point ground is widely and successfully used by telephone companies for their digital switching facilities, their required maintenance is a real concern. The worst possible grounding system is to have a single-point ground area with one additional grounding connection at a remote point. This configuration would provide a well-defined current path through the area requiring the highest degree of protection. Another reason not to use single-point grounding for the entire SRG is the necessity of supplying large areas with several sources of power, each with its own ground connections. These widely separated ground connections, when ultimately tied together, form large unwanted ground loops, which are another source of electrical noise.

9.10.13.3 Summary of Recommended Practices for SRGs and Their Installation

(1) Follow applicable codes and standards for safe grounding. There is no conflict between safe grounding for people and effective HF grounding for sensitive electronic equipment.
(2) Select a suitable SRG approach and assure that it is installed and maintained properly.
(3) Permanently bond the SRG to all accessible building steel and to each metallic path crossing the plane of, or within 6 ft of, the SRG.
(4) If a single point of entry for power and grounding cables into the space exists, then single-point grounding of the area to local building steel is acceptable if this grounding system is verified periodically by skilled test personnel. The sensitive equipment is multipoint grounded to the SRG.
(5) Bond the SRG to each piece of sensitive equipment.
(6) Bonding connections to the SRG should be as short as practical with no sharp folds or bends.
(7) If more than one bonding conductor is used, they should be connected to separate parts of the equipment and to separate points on the SRG.
(8) Connections of sensitive electronic equipment to the SRG should not be made on the outermost grid conductor. Heating, ventilation, and air conditioning equipment and panelboards can be connected to the outermost grid conductor. Critical equipment should be located and bonded to the SRG >6 ft away from building steel or other potential lightning paths.
(9) All separately derived systems serving equipment located on the SRG should have their power grounding-point (i.e., neutral-to-ground bond) bonded to the SRG.
(10) All cooling, heating, ventilation, and air conditioning equipment, associated piping, metal wall studs, panelboards, switchboards, transformers, and similar equipment within the protected area, shall be bonded to the SRG.
(11) No connections should be made to remote or dedicated earth grounding points or any similar attempt at separate earth ground paths.
(12) All data and power cables should lay on or very close to the SRG.
(13) Documentation should be complete in all details, including the proper grounding and bonding of heating, ventilation, and air conditioning equipment, piping, raceways, and similar items. The responsible engineer should not expect installers to complete the design.

9.10.13.4 Bonding Straps from Equipment to the SRG. Sensitive electronic equipment should be connected to the SRG with low-inductance conductors. Flat foil strips are the recommended practice. Connections to the equipment frame or OEM supplied terminal are critical. Paint or other surface contact inhibitors should be removed before bonding straps are attached. Straps should be as short as possible to minimize inductive reactance. The use of two bonds is recommended to further reduce reactance. The straps should not be folded or coiled to radii of less than 8 in for best performance. Even in equipment lineups where the equipment is bolted together, the recommended practice is to bond each enclosure to the SRG with its own strap.

9.10.13.5 SRS for Noncontiguous Areas. The SRG is appropriate for contiguous areas but is impractical and can be ineffective between widely separated areas or buildings. Solid metal cable tray or conduit (bonded between sections) should be used to duct signal or power cables between areas of intensive sensitive equipment usage. These cable trays or conduits should be grounded at both ends and at several points along their length. Recommended practice is to also augment these forms of SRS with TVSS devices. Other methods (e.g., optical isolators or balun transformers) can also provide increased noise and surge immunity for the interconnected data circuits.

9.10.14 Earth Grounding Electrode Resistance and Impedance. Recommended practice is to provide a resistance of 25 Ω or less for any made earth grounding electrode, but one does not need to seek unrealistically low values of resistance such as the popular 1 Ω value. Such connections are only used for electrical safety and are not suitable for use at HF. Recommended practice is to follow the guidelines given in IEEE Std 142-1982 [12].

The use of earth grounding electrodes as a useful means of conducting HF currents is effectively refuted as a design goal when one reviews the relationship of earthing impedance with increasing frequency. At frequencies above a few tens of kilohertz, earth grounding electrode impedances are considerably larger than at dc or 60 Hz. Measured resistive values of 1 Ω or 5 Ω are only meaningful at dc and possibly at low frequencies. In any case, these are not HF impedances; and the associated frequencies where the ohmic values may be low are not considered to be typical noise frequencies in relation to digital electronic systems, which may require effective grounding from dc to 25–30 MHz or sometimes higher, i.e., HF.

9.10.15 Earth Grounding Electrode Conductor Resistance and Impedance. The impedances of typical grounding electrode conductor wire sizes linearly increase as a function of frequency. This further frustrates a designer's efforts to achieve a low impedance for the earth grounding electrode at high frequencies.

9.10.16 Recommended Grounding of Communication Systems. It is important to ensure that low-impedance grounding and bonding connections exist among the telephone and data equipment, the ac power system's electrical safety-grounding system, and the building grounding electrode system. This recommendation is in addition to any made earth grounding electrodes, such as the lightning ring-ground. Failure to observe any part of this grounding requirement may result in hazardous potentials being developed between the telephone (data) equipment and other grounded items that personnel may be near or might simultaneously contact.

To accomplish the needed grounding, the telephone (data) equipment should be grounded to nearby building structural steel or the metallic, cold-water piping system. The ac/dc power supply for the telephone (data) equipment should be locally grounded via a low-inductance grounding jumper. This recommendation is in addition to the equipment grounding conductor in the associated branch circuit and ac supply cord for the telephone equipment ac/dc power supply. If the central office trunk cable (COTC) is terminated in the equipment area, its outer shield and associated TVSS should also be bonded to all of the above items in a low-impedance manner. This approach conforms to telephone industry[26] practice. The outside plant cabling terminations for cable accessed television (CATV) should be similarly grounded.

If the telephone COTC is terminated on a plywood backboard, equipped with a SRP on its face, i.e., #22 Ga. galvanized steel, then the recommended grounding can be accomplished by connecting any point on the SRP to the other indicated grounding and bonding points (e.g., building steel, grounded water pipe). Low-inductance grounding and bonding means should be used to make the connections. A similar arrangement may exist on each floor of the facility where the telephone (data) system is accessed in a wiring closet, and wiring terminals (punchdown blocks) are available for connection into the instruments on that floor from the telephone (data) *risers*.

9.10.17 AC System Grounding for Uninterruptible Power Supply (UPS) With Bypass Circuits. A discussion of the various recommended and nonrecommended means of achieving interconnection of *prime* and *backup* ac supply sources via transfer switches (any kind) is presented in detail in reference IEEE Std 446-1987 [14]. This reference also discusses uninterruptible power system installation requirements.

9.10.18 Grounding Air Terminals (Lightning Rods). Air terminals are recommended to be treated in similar fashion as the buried ring-ground and as any earth grounding-electrode (NEC [2], ANSI/NFPA 780-1992 [4]). Interconnection of the air terminals with the overall premises grounding system is recommended practice, which assists in minimizing potential differences, destructive arcing, and the associated problems of common-mode electrical-noise currents appearing via the grounding system.

9.10.19 Galvanized Construction Channel as a Grounding Bus-Bar. Galvanized construction channel is generally recommended to mechanically support and to secure items in place. If installed properly this channel makes a good HF electrical grounding bus-bar *and* connection point for conduits, piping, or similar items that are required to be bonded to one another. In this application, construction channel is generally an effective conductor for frequencies up to tens of megahertz.

---

[26] Formerly *Bell*.

9.10.20 Uninsulated Grounding Conductors. The use of uninsulated (bare) conductors is not recommended in any manner except when used for short grounding jumpers, bonding jumpers, and similar items that are not enclosed in conduit/raceway. Insulated wire is recommended for all forms of circuits, grounded or ungrounded, that are contained within equipment or conduit/raceway systems.

The use of uninsulated (bare) wire for the EGC within a conduit/raceway is not recommended. Application of uninsulated (bare) conductors within a metallic conduit/raceway may cause two unwanted conditions:
(1) Localized destructive arcing between the uninsulated (bare) conductor and the metallic conduit/raceway along the uninsulated conductor's path during ground fault or lightning conditions that damage other insulated conductors, or
(2) Generation of low-level HF currents (electrical noise) due to intermittent contact with the metallic conduit/raceway.

9.11 Lightning/Surge Protection. ANSI/NFPA 780-1992 [4] provides both an isokeraunic map of thunderstorm days and a Risk Index calculation procedure. These procedures provide background data for risk decisions regarding lightning protection. Based upon the nature of the sensitive equipment, its high cost of repair or replacement, and general value to operations, a lightning protection program modeled on ANSI/NFPA 780-1992 is advisable for most sites.

In general, only lightning protection components listed to ANSI/UL 96-1985 [5] should be used. Recommended practice is that the lightning protection system be designed, constructed, and installed in conformance with UL 96A-1982 [16]. A structure or building so equipped may be master labeled for structural lightning protection by UL or an acceptable equivalent, and should be so labeled as evidence of proper protection.

UL 96A-1982 [16] provides for reconditioning of existing installations previously conforming to the standard's requirements. This procedure should be followed to maintain the labeled status of the installation.

In addition to basic structural lightning protection means, installation of a listed secondary lightning arrester is recommended at the service entrance of all major sensitive electronic equipment facilities. A listed and properly rated TVSS should be applied to each individual or set of electrical conductors (e.g., power, voice, data) penetrating any of the six sides forming a structure. All power-circuit TVSS devices should be coordinated per IEEE Std C62.45-1987 [10]. All signal-circuit TVSS devices (primary and secondary surge protectors) should be per ANSI/EIA/TIA 571-1991 [1], IEEE Std C62.36-1991 [8], UL 497-1991 [17], and UL 497A-1990 [18].

9.11.1 Service Entrance Lightning/Surge Protection. Facilities housing major electronic computer/data processing systems or other major sensitive electronic equipment should have service entrances equipped with effective lightning protection in the form of listed secondary surge arrestors. Care should be taken to assure that the method used for the installation of TVSS equipment will not cause a degradation of its current-diverting and voltage-clamping abilities.

It is also recommended that if gas tube surge arrestors are used, there should be a paralleled installation of wavefront modification capacitor(s) for the purpose of reducing HF components in the leading edge of the lightning surge current's waveform, and for acting as a HF bypass capacitor for HF currents whose associated potentials are below the operating thresholds of the arrester.

9.11.2 Premise Electrical System Lightning/Surge Protection. In addition to secondary-rated lightning arrestors in the service entrance equipment, it is recommended that additional TVSS be applied to downstream electrical switchboards and panelboards if they support electronic load equipment (see Fig 9-6). The additional TVSS devices are recommended to be sized and installed per IEEE Std C62.45-1987 [10] requirements to achieve proper coordination.

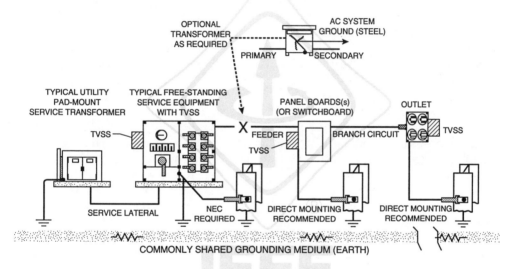

**Fig 9-6
Typical Locations of Power Distribution TVSS**

TVSS used for ac circuit protection is generally recommended to be connected in all combinations of line-to-line, line-to-neutral, line-to-ground, and neutral-to-ground. TVSS devices should be listed to UL 1449-1985 [19]. Recommended TVSS installation practice is for all lead-lengths to be short and shaped to minimize loop-geometry, such as by twisting all the phase, neutral, and grounding conductors together and by avoiding any sharp bends in the conductors. IEEE Std C62.41-1991 [9] and UL 1449-1985 [19] should be used as a standard means of verifying performance of TVSS devices. However, they will not perform properly under field conditions of use unless installed in a correct manner.

9.11.3 UPS Surge Protection. Lightning and other transient voltage producing phenomena are harmful to most UPS equipment and to sensitive electronic load equipment (e.g., via an unprotected static-switch bypass path around a UPS). Therefore, it is recommended practice that both the rectifier-charger input circuit to the UPS and the associated UPS bypass circuits (including the manual maintenance bypass circuit) be equipped with effective Category "B" TVSS protection as specified in IEEE Std C62.41-1991 [9].

The required protection should be attached in both the line-to-line and line-to-ground modes. Low-inductance connections should be employed for this protection. TVSS devices should be connected to the available equipment grounding conductors, not to isolated grounding conductors.

9.11.4 Data (Communications and Control) Cabling and Equipment Lightning/Surge Protection. Typically, data and control system cabling are routed throughout a facility. Sensitive equipment can be connected at both cable ends, and is also locally connected to ac power circuits. Typical equipment might be video display terminals (VDTs) and personal computers (PCs). The cables run various distances between units of the overall system, producing conditions where lightning currents can enter the equipment from either the ac supply connection or the data cabling. Both conducted and coupled noise paths are generally possible.

Since many forms of electrical interference (e.g., lightning or switching induced surges) may be coupled into data-communications cabling, it is recommended practice that data cables are installed with minimum open-loop areas. This practice is especially important in large facilities where such cables are often routed over long distances. Such routing often creates large open-loop areas and, hence, maximum coupling between an interference source and the victim cable(s). Containing data cables in fully enclosed metal conduit/raceway and otherwise routing them along structural surfaces that are grounded is also a recommended means of minimizing the coupling problem.

It is recommended that each ac supply connection for equipment, such as a VDT, which connects to both ac power and to data cabling, be equipped with a locally installed TVSS at the branch circuit outlet where ac power is taken to operate the equipment. These TVSS devices should generally provide Category "A" protection, as specified in IEEE Std C62.41-1991 [9]. TVSS devices should be referenced to the applications and test conditions described in IEEE Std C62.41-1991, with the maximum let-through voltages defined for specific test waveforms and current levels.

Data (e.g., control, data I/O cables, and other communication or signaling) cables that are metallic (not fiber optic) should also be equipped with suitable TVSS (surge protector) devices. The surge protector devices should be specifically tailored to the type of signal being used on the cable (e.g., RS-232C, RS-422). Surge protectors should be placed on both ends of data cables for optimum effect. However, the electronics end of the cable alone may be protected as a minimum precaution if the far end does not connect into electronics, but into passive elements such as a contactor coil, indicator lamps, or dry contacts.

Surge protectors that exhibit large capacitances are not generally suitable for connection across data circuits with fast rise or fail time pulses, since the surge protector capacitance can overly distort the desired signal. The general recommendation for data circuits is use of a hybrid form of surge protector, which generally consists of a gas-tube on the unprotected side of the line, followed by a series impedance, followed by a fast-acting silicon avalanche diode as a shunt element. Both balanced and unbalanced hybrid TVSS are available. They can be used over a broad range of voltages and data rates.

Installation of (power circuit) TVSS devices and (signal circuit) surge protectors typically entail mounting them directly to the metal frame/enclosure of the equipment for effective operation. Recommended installation practices are provided in the following section.

9.11.5 Telecommunication System Lightning/Surge Protection.

Most facilities have telecommunication systems that are often used for both voice and data transmission. Both separate and combined (copper-based) voice/data wiring systems need protection from lightning surges. Effort should be undertaken to ensure that the telecommunications system is properly grounded and surge-voltage protected. (Telephone industry terminology for a TVSS device is *surge protector*. Surge protectors employed at building entrances are termed *telecommunications protector or station protectors*, and surge protectors employed within the building or terminal equipment are termed *secondary protectors*.) Surge protectors for telecommunication circuits should be listed under UL 497-1991 [17] or UL 497A-1990 [18]. The chosen surge protector should be designed and tested for the intended type of telecommunication circuit.

9.11.5.1 Central Office Trunk Cables.

A lightning/surge protection system should be provided at the location at which telephone COTCs enter a building, depending on the local probabilities of transient disturbances. This TVSS protection is normally provided by the telephone company (public utility) as a part of their installation.

9.11.5.2 Main Distribution Frame and Intermediate Distribution Frame Terminations.

Not all troublesome surge currents enter sensitive equipment via the COTC. An alternate mode of entry is via the premises telecommunications wiring system. The premises telecommunications wiring system may span multiple buildings, as well as multiple locations within each building. The telecommunications wiring system is often sufficiently spread throughout the facility to develop large common-mode potentials from its origination to its termination points. Therefore, telecommunication cables often need supplementary surge protectors at intermediate termination points and at the instrument-end of the cables. These surge protectors should be installed between each conductor (in a pair) and a local (low-inductance) grounding means.

9.11.5.3 Referencing Telecommunications TVSS Devices. Plywood backboards, used to mount signal-wiring terminal blocks, secondary surge protectors, and TVSS devices for related power circuits, should be covered with #22 Ga. galvanized sheet steel to form a SRP as shown in Fig 9-7. Alternatively, surge protective devices should be housed in a metallic enclosure of equal functionality as an SRP. All telecommunications cable grounds and associated surge protectors should be directly referenced to this SRP. The foregoing is contrasted with the typical use of nonmetal sheathed plywood that forces the use of long, highly inductive, discrete grounding conductors to connect the TVSS devices to ground. These long grounding conductors increase the TVSS clamping-voltage, due to the classical effects of rapidly changing current through an inductance, i.e., $E_t = L\, di/dt$.

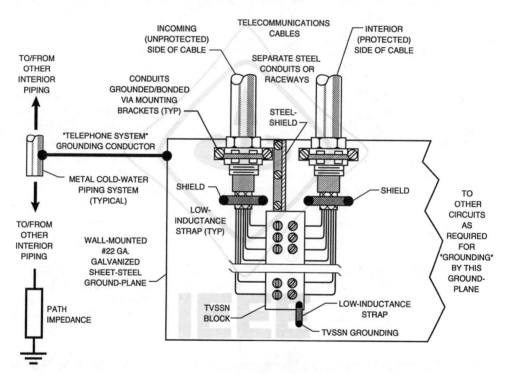

Fig 9-7
Recommended Installation Approach for Signal Protectors

TVSS devices that protect ac/dc power supplies, or similar devices that are part of a telecommunication system, should be equipped with appropriate TVSS devices/surge protectors and referenced to the same SRP as the communications circuit TVSS devices are referenced, as shown in Fig 9-8.

9.11.5.4 Surge Reference Equalizers. With the expanding use of smart electronics that have a power port connection as well as a communications port

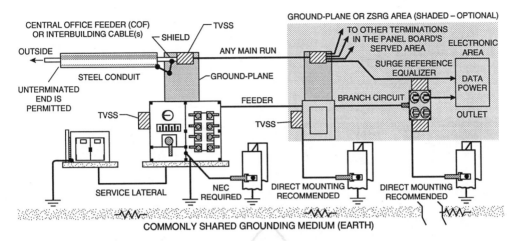

Fig 9-8
Recommended Installation Practice for
Combined TVSS and Signal Protectors

connection (telephone answering machines, fax, desktop publishing, industrial process control, remote terminals, etc.), a new risk of damage has emerged. Surge protection might have been provided on both the power line and the communications line, yet the equipment could be damaged by a difference of reference voltages developed during a surge event. One scenario involves the voltage difference between the two chassis (and thus the signal reference) of subunits powered from different branch circuits, where a surge-protective device operation changes the voltage with respect to the equipment grounding conductor of one subunit (Fig 9-9) [B2]. The other scenario involves the difference of voltage resulting from the operation of a surge protective device at one of the entrances of the power or communications lines (Fig 9-10).

A potential solution for this type of problem is to provide a fiber optic data link rather than a metallic connection. When this approach is not practical, another remedy to this situation is a surge reference equalizer, a packaged assembly intended to mitigate the threats associated with surges carried by the conductors of either, or both, the power system and the communications system, as well as differences in reference voltages between the two systems. The device is installed next to the equipment to be protected and combines the protective function of both systems in the same enclosure. The device is plugged in a power receptacle near the equipment to be protected, with the communications system wires (telephone or data link) or the coaxial cable (TV) routed through the enclosure. A common, single grounding connection equalizes the voltages of the two "grounding" conductors through the grounding prong of the three-prong power line plug.

Large surges on the power system originating outside of the user's facility, associated with lightning or major power-system events are best diverted at the service entrance of a facility. While such a protection is not mandated at present, trends indicate that a growing number of application documents

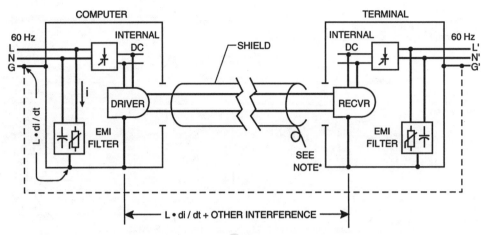

NOTE: Some organizations insist on grounding one end only of the shield.

**Fig 9-9
Signal Reference Potential Difference Created by Diverting Transient
Energy Through Grounding Conductor of Portion of System**

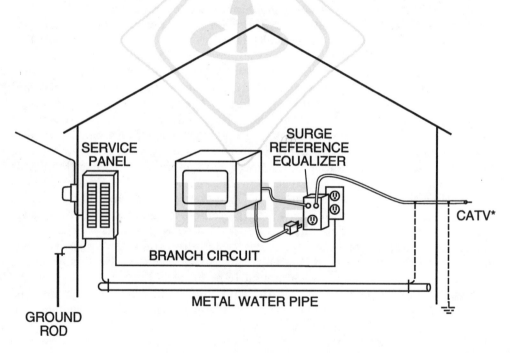

NOTE: Some local authorities require an external grounding connection; some others require an internal bond as shown.

**Fig 9-10
Protection Against Reference Voltage Difference Caused By
Two-Port Connection**

support a recommendation for this protection. Surges generated within the premises can be diverted by protectors located close to the internal surge source or close to the sensitive equipment, such as reference equalizers. This possibility of dual protection raises the issue of coordination of cascaded devices, an emerging concern in the application of surge-protective devices in the power system of end-user facilities [B1].

9.11.5.5 Exterior Building Systems and Piping Lightning/Surge Protection. All exterior mechanical system items (e.g., cooling towers, fans, blowers, compressors, pumps, and motors) that are in an area not effectively protected by a lightning protection system per ANSI/NFPA 780-1992 [4] should be considered as targets for a lightning strike. Therefore, it is recommended practice to individually provide TVSS protection on both the power input and data circuits of all such electrical equipment. This TVSS should be Category "B" or "C" devices (as specified in IEEE C62.41-1991 [9]), depending on building location and system reliability requirements.

Galvanized construction channel is typically recommended to be used to support piping and conduits. Due to its geometry, this material makes a low-inductance grounding bus for the interconnection of pipes and conduits to one another and to building structural steel, or equivalent. This material is also good for structural supporting, grounding, and bonding of electrical equipment enclosures. However, wire-conductors used to connect the channel to other items can create a high-inductance link. Therefore, the channel is best used by itself as a directly mounted bus-bar.

Any metal pipe or conduit (exposed conductor) that runs externally to the building and then also extends back into the building (especially if the extension is into an electronic load equipment area, such as the piping for heating, ventilation, and air conditioning) has a real possibility of the external portion of the item being struck by lightning. It is capable of carrying a lightning voltage and current back into the building and arc, i.e., side-flash, from the energized item to other grounded items. This concern is real from both a shock and a fire hazard standpoint. Therefore, all such metallic items should be grounded to the building structural steel as they pass in/out of the building. Bonding of all such pipes, electrical conduits, and similar items into a single electrically conductive mass is very important. If nearby structural steel is not available, all items should be bonded to the local electrical system and to the lightning ring-ground via a down-conductor system generally installed as a lightning conductor per ANSI/NFPA 780-1992 [4].

9.12 400 Hz (380-480 Hz) Power Systems. Certain sensitive electronic loads require 400 Hz power. The design of 400 Hz power systems requires additional considerations beyond those for 60 Hz power. Thus, 400 Hz power is on-site generated or converted from a 60 Hz supply by an engine-alternator/generator, a motor-alternator/generator set, or solid-state frequency converters. Since these supplies generally may not be solidly interconnected to a 60 Hz ac system, they should be treated as SDS designs.

9.12.1 Recommended Location of the 400 Hz AC System. 400 Hz systems are recommended to be located near their served loads due to concerns over common-mode noise currents and 400 Hz wiring voltage drops. 400 Hz sources are also recommended to be located on (and bonded to) the same SRS as the related 60 Hz ac system and their served loads.

9.12.2 General Grounding and Shielding of 400 Hz Systems. Generally, on 400 Hz ac systems (most of which are three-phase) the neutral point of the ac supply should be SG in accordance with applicable codes and standards. This design allows return current paths for any LC filters in the load equipment. Grounding and bonding practices for 400 Hz systems and equipment are essentially the same as for the 60 Hz systems and equipment, with the sole exception of not being able to make 400 Hz ac supplies solidly interconnected systems with respect to their 60 Hz supply.

9.12.3 Controlling 400 Hz Wiring Losses. Wiring conductors exhibit significantly greater inductive reactance and ac resistance (skin effect) at 400 Hz than at 60 Hz. Typical wiring impedances for single copper conductors are shown in Table 9-1. Nonferrous metallic raceway is recommended for 400 Hz conductors, which significantly reduces the losses resulting from ferrous metal conduit/raceway. Aluminum conduit/raceway is generally recommended practice. Nonmetallic raceways should not be used due to their lack of electrostatic shielding. If 400 Hz power must be provided to loads over distances more than a few tens of feet, wire size and line drop compensators (LDCs) should be evaluated for controlling voltage losses. Increasing wire size above #1/0 AWG does not offer significant reductions of wiring impedance. The use of paralleled conductors can be effective for reducing wiring impedance. When using paralleled conductors per phase as a means of controlling voltage drop in 400 Hz feeders or branch circuits, the paralleled conductors are better carried in separate conduits or raceways which are made up as mirror (identical) images of one another. Paralleled, separate conduit/raceways effectively reduce the ac resistance and inductance of the 400 Hz circuits. The use of parallel conductors per phase in the same conduit or raceway reduces the ac resistance but not the inductive reactance (which is the dominant component of the wiring impedance at 400 Hz).

9.12.3.1 400 Hz Line-Drop Compensators. Passive or active LDCs are often applied to 400 Hz circuits as a means of reducing wiring voltage drop and improving the voltage regulation on the circuit.

Passive LDCs provide capacitive reactance to cancel the wiring inductive reactance. Passive LDCs exhibit unwanted frequency-dependent characteristics that reduce their effectiveness with harmonic-rich load currents. Passive LDCs must be properly located in the circuit to be effective. The location of the passive LDC can be anywhere along the circuit where its inductive reactance is desired to be canceled. For example, with a line feeder connected to a group of short branch circuits, the LDC can be placed anywhere along the feeder ahead of the point where the split is made to the branch circuits. When a short

Table 9-1
415 Hz Impedance in Ohms/100 ft
(single copper conductors)

Wire Size (AWG)	In Air			Non-Metallic Conduit			Rigid Aluminum Conduit			Rigid Steel Conduit		
	Rac	X1	Z	Rac	X1	Z	Rac	X1	Z	Rac	X1	Z
8	0.0782	0.0267	0.0826	0.0782	0.0321	0.0845	0.0782	0.0321	0.0845	0.0784	0.0401	0.0880
6	0.0493	0.0258	0.0556	0.0493	0.0310	0.0582	0.0493	0.0310	0.0582	0.0497	0.0387	0.0630
4	0.0315	0.0248	0.0401	0.0315	0.0297	0.0433	0.0315	0.0297	0.0433	0.0322	0.0371	0.0492
2	0.0198	0.0235	0.0307	0.0198	0.0284	0.0346	0.0198	0.0284	0.0346	0.0202	0.0354	0.0408
1	0.0164	0.0232	0.0284	0.0164	0.0279	0.0324	0.0164	0.0279	0.0324	0.0175	0.0348	0.0390
1/0	0.0135	0.0229	0.0266	0.0135	0.0275	0.0307	0.0135	0.0275	0.0307	0.0152	0.0344	0.0376
2/0	0.0115	0.0224	0.0252	0.0115	0.0269	0.0293	0.0115	0.0269	0.0293	0.0135	0.0336	0.0362
3/0	0.0097	0.0220	0.0240	0.0097	0.0264	0.0281	0.0097	0.0264	0.0281	0.0118	0.0330	0.0350
4/0	0.0084	0.0218	0.0233	0.0084	0.0261	0.0274	0.0084	0.0261	0.0274	0.0110	0.0328	0.0346
250 MCM	0.0076	0.0217	0.0230	0.0076	0.0260	0.0271	0.0076	0.0260	0.0271	0.0101	0.0325	0.0340
300 MCM	0.0070	0.0212	0.0223	0.0070	0.0255	0.0265	0.0070	0.0255	0.0265	0.0097	0.0319	0.0333
350 MCM	0.0064	0.0212	0.0221	0.0064	0.0254	0.0262	0.0064	0.0254	0.0262	0.0093	0.0318	0.0331
400 MCM	0.0061	0.0211	0.0219	0.0061	0.0253	0.0260	0.0061	0.0253	0.0260	0.0091	0.0315	0.0328
500 MCM	0.0054	0.0206	0.0213	0.0054	0.0248	0.0253	0.0054	0.0248	0.0253	0.0084	0.0310	0.0321

feeder is used with several long branch circuits it is best to apply individual LDCs to each of the branch circuits and not on the shared feeder. Thus, if a branch circuit is switched on/off it has minimal effect on the system voltage.

Active LDCs regulate the source output voltage in response to a remote voltage sensing point (such as near the load) or in proportion to the load current. OEMs with 400 Hz converters are generally the suppliers of LDCs that are matched to the product and application to ensure compatibility. It is recommended that LDCs be used only if voltage drop cannot otherwise be practically controlled by locating the 400 Hz source close to the loads or through the use of parallel conductors.

9.12.4 400 Hz Conductor Ampacity. The current-carrying capability of a given conductor is less at 400 Hz than at 60 Hz, due to additional stray losses (primarily skin effect). The derating factor for 400 Hz conductor ampacity, based on the 60 Hz conductor ampacity, is as follows (NEC [2]):

$$\text{derating factor} = \sqrt{\frac{1}{R_{ac}/R_{dc}}} \qquad \text{(Eq 9-1)}$$

where
  R_{ac} = ac resistance of the conductor at 400 Hz
  R_{dc} = dc resistance of the conductor

When more than three current-carrying conductors are installed within the same raceway, most applicable codes and standards require that ampacity derating factors be applied. For most applications this factor is 80% of the ampacity resulting from Eq 9-1. Further reductions may be required for ambient temperature and total number of wires sharing the same raceway.

9.12.5 Component Derating at 400 Hz. When 60 Hz components are used in 400 Hz applications, component derating is often required. Selecting components for 400 Hz applications is difficult because no national standards exist for electrical items to be tested and listed at 400 Hz. Therefore, 400 Hz installations should be carefully inspected. Manufacturers should be contacted for application assistance.

Overcurrent protective devices may require special derating at 400 Hz. Fuses are typically not appreciably affected by 400 Hz power, but thermal-magnetic and magnetic-only circuit breakers are affected. When circuit breakers are used at 400 Hz, they should be sized for the expected load using the derating factors supplied by the OEM. Circuit breakers of 60 Hz generally do not possess the same trip-calibration curve nor interrupting capacity at 400 Hz as at 60 Hz. Once a 60 Hz circuit breaker is applied to the 400 Hz system and is derated, the OEM embossed or the permanent label for 60 Hz ampacity may still be visible. This incorrect label is confusing and may cause problems. Increased attention to field-applied labels and advisory signs is recommended as well as a specific ruling from the local electrical safety inspection authority having jurisdiction at the location. Where possible, components should be used that bear the OEM's 400 Hz ratings.

9.13 Switchboards. Switchboards serving feeders that support sensitive electronic equipment and related sensitive loads are discussed below.

9.13.1 Grounding Bus-Bars. Switchboards should always be equipped with a complete EGC bus-bar system. The need for a bus to terminate EGCs is well-established, as almost every feeder that supports sensitive electronic equipment will require an EGC. Termination of these EGC wires without a proper bus degrades the reliability of the grounding path, especially for HF currents.

9.13.2 Neutral Bus-Bars. Neutral bus-bars in switchboards may be rated less than the line current bus-bars due to diversity factors. Ratings in the 80% range are not uncommon, although ratings as low as 50% may be seen. Such derating can be a problem if the switchboard directly serves line-to-neutral connected loads, as discussed in the following paragraph.

Switchboards employing neutral buses for three-phase, 208Y/120 Vac, 4-wire + ground configurations, may encounter high harmonic currents on the bus due to multiple line-to-neutral connected, nonlinear loads (e.g., capacitor-input rectifier circuits). This condition relates principally to third-harmonic currents and generally is the basis for oversizing the switchboard neutral bus.

Harmonic current is a concern on switchboards only when the output feeders are connected directly to the harmonic-current producing loads in a line-to-neutral fashion. Switchboards often supply line-to-line connected loads or delta-wye connected isolation transformers. The delta-wye transformation has the ability to reduce third-harmonic current on its primary via the action of the delta-connected winding. Delta-wye transformer usage to feed harmonic-current producing loads is a recommended practice.

Neutral bus-bar assemblies may be specified by the switchboard OEM with a heavy duty rating without affecting the product safety listing. This approach allows a 2× rated neutral bus-bar to be placed into a 1.25× rated line bus-bar switchboard so as to allow for high third-harmonic current conditions without requiring a 2× line bus-bar rating to get a 2× rated neutral bus-bar. For example, a 1200 A switchboard is not required to obtain the needed 1200 A neutral bus-bar rating since a special, listed 600 A switchboard may be obtained with a 1200 A neutral bus-bar. This approach is recommended practice where excessive neutral current is anticipated.

9.14 Panelboards. Panelboards used to support sensitive electronic equipment should be the industrially rated style for power or lighting applications, and should not be a lighter-duty type.

9.14.1 Recommended Line Bus-Bar Ampacity. Minimum line bus-bar ampacity should be calculated on the basis of maximum present and future, full-load, rated, true rms current.

9.14.2 Recommended Neutral Bus-Bar Ampacity and Wiring Capacity. Where a neutral bus-bar assembly is included in a three-phase panelboard and the panelboard is expected to support sensitive electronic loads with high third-harmonic currents resulting from line-to-neutral connected nonlinear loads [B4], the neutral bus-bar assembly should be upsized in relation to the line current bus-bars by a factor of no less than 1.73. Listed panelboards with a 2× upsizing of their neutral bus-bar assembly are recommended.

The wiring capacity of the neutral bus-bar is important when used to support sensitive electronic load equipment. Many of these loads are line-to-neutral connected. Individual (dedicated), neutral conductors should be used, and individual termination points should be available on the neutral bus-bar for each possible load using a line-to-neutral connection.

9.14.3 Equipment Ground Bus-Bar Ampacity and Wiring Capacity. The neutral and equipment grounding bus-bars should be separate and jumpered together only when appropriate, e.g., when used at the service entrance or sometimes after a separately derived source. The EGC bus-bar in a panelboard is not required to have an increased ampacity to support sensitive electronic loads. Harmonic currents such as may be found on neutral bus-bars are not a problem in this regard. The ground bus-bar assembly should be listed for inclusion in the panelboard of interest.

Wiring capacity of the equipment grounding bus-bar is important for panelboards that supply sensitive electronic equipment loads since all loads normally require an individual, dedicated EGC. There should be an individual termination point available on the EGC bus-bar for each possible load. There may be additional requirements for termination of EGCs if IG is employed on the served circuits. The additional requirements can double the number of EGCs to be terminated to the EGC bus-bar in the panelboard.

9.14.4 Panelboard Mounting and Grounding. Panelboards should be directly mounted to any local (grounded) structural building steel member.

NOTE: Attachment of a panelboard to well-grounded structural building steel is not generally considered to be a cause of grounding problems involving HF common-mode noise. An exception would be a case in which other unwanted grounding and bonding conditions exist on the installation, which are the actual root cause of the noise problem.

Isolation of a panelboard from structural building steel by an electrically insulating material is not recommended practice, since considerable HF current may be coupled between two closely positioned metallic objects. ANSI/NFPA 780-1992 [4] requires good grounding and bonding between objects such as structural building steel and a panelboard if within side-flash distance (approximately 6 ft, horizontally) of each other. Insulation materials, commonly used to separate a panelboard from structural building steel, are rarely capable of withstanding lightning-induced arcing conditions.

9.14.5 Location of the Panelboard. Panelboards that serve sensitive electronic equipment should be placed in the same area as the load equipment, and should be bonded to the same SRS as their served loads. This location philosophy is recommended for any panelboard that serves loads in the same area with the sensitive electronic equipment, such as lighting and heating, ventilation, air conditioning, and process cooling equipment. Downstream (sub-) panelboards that are located in other sensitive electronic equipment areas should also be bonded to the respective SRS of the served loads.

9.15 Power Distribution Units (PDUs) for AC Power-Load Interface. The PDU is recommended as the principal means of supplying ac power and the grounding interface between the premises electrical system and connected sensitive electronic load equipment. It is generally a superior method to almost all available building wiring techniques. A listed PDU is essentially a prefabricated ac power and grounding system, including flexible output cables, for powering sensitive electronic equipment. The PDU is also referred to as a computer power center (CPC).

PDUs may contain an automatic line voltage regulating transformer (ALVRT), motor-alternator/generator (M-G), or even full UPS capability, instead of only containing a simple IT. Combinations of these items are also possible. Other forms of PDU used for special applications may be constructed without an internal means of isolation or transformation. These PDUs should be used with an externally provided IT, M-G set, or UPS that is a part of the premises wiring system for proper operation.

9.16 Automatic Line Voltage Regulating Transformers (ALVRTs). When the nominal supply voltage is not stable, application of a *carefully chosen* ALVRT can provide the necessary voltage correction. The ALVRT is recommended to be placed near its served load equipment, and generally should be configured as a SDS. The use of an ALVRT as a noise filter is not generally recommended.

9.17 Dry-Type Transformer. Dry-type shielded isolation-transformers are recommended to be used near their served loads. They provide system voltage matching and create a SDS physically near their connected electronic load equipment. These transformers should be the shielded isolation-type and should not be autotransformers.

9.17.1 Electrostatic Shielding in the Transformer. IT used to support sensitive electronic load equipment should be equipped with at least a single-layer electrostatic shield of the primary-secondary interwinding type. The shield should be grounded or bonded to the transformer metal frame/enclosure using low-inductance means. A second electrostatic shield, referenced to the primary electrical-line voltage, is also recommended. This construction should be used in most cases to reduce common-mode HF currents (electrical noise) via conversion to normal-mode electrical noise.

9.17.2 Nonlinear Load Impact on Transformers. Nonlinear loads may cause overheating in both liquid-immersed and dry-type transformers. In such cases recommended practice is to derate the conventional transformers, and where practical, to specify K-factor rated transformers that are specifically designed for nonsinusoidal loads.

9.17.2.1 Derating Conventional Transformers. Conventional power transformers usually require derating when serving nonlinear, typically electronic, loads. To obtain rated transformer performance an approximately sinusoidal and balanced line voltage and load current are required. When the load current is distorted, an appropriate derating can be calculated using formulas provided in 4.4.4.1 of IEEE Std C57.110-1986 [7]. There are two possible methods: a detailed measurement, and a simplified measurement plus calculation.

Another derating method sometimes suggested in the computer industry compares only the crest factor of the load current to the crest factor of a sinusoidal waveform. The transformer derating would then be equal to the sinusoidal crest factor divided by the actual or predicted crest factor of the load current. In the case of high third harmonic content (the case of capacitor-input dc power supplies), this method may be a reasonable, albeit imprecise rule of thumb. This method may underestimate losses in the presence of harmonics of higher order, and it does not take into consideration differences in the losses associated with rated eddy current in the transformer.

Figure 9-11 shows an example of a derating curve, appearing in [B4], obtained by more accurate computations based on IEEE Std C57.110-1986 [7]. This figure shows that the derating can reach 50% when the transformer supplies more than 70% of its load to the capacitor-input dc power supplies of very distorted electronic loads. This example shows the importance of performing the calculations in accordance with IEEE Std C57.110-1986 when selecting a transformer supplying electronic loads for a new installation, or in an existing installation where the percentage of electronic loads is increasing.

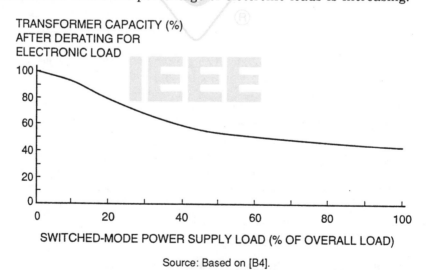

Source: Based on [B4].

Fig 9-11
Transformer Capability for Supplying Electronic Loads

9.17.2.2 K-Factor Rated Transformers. UL and transformer manufacturers have established a K-factor rating for dry-type power transformers to indicate their suitability for nonsinusoidal current loads. The K-factor relates a transformer's capability to serve varying degrees of nonlinear load without exceeding the rated temperature-rise limits. For new installations or for replacements, K-factor rated dry-type transformers are recommended. The number of manufacturers and available kVA ratings is expected to increase with increasing use of nonlinear electronic loads.

Standard K-factor ratings are 4, 9, 13, 20, 30, 40, and 50. The K-factor for a linear load is one. For any given nonlinear load, if the harmonic current components are known, the K-factor can be calculated and compared to the transformer's nameplate K-factor. As long as the load K-factor is equal to or less than the transformer's rated K-factor, the transformer does not need to be derated.

A K-factor rated transformer is preferred over an oversized (derated) conventional transformer for several reasons. The K-factor transformer is equipped with 200% rated neutral bus and is likely to be smaller and less expensive. An oversized transformer may require a higher short-circuit rating on circuit breakers and may draw a higher inrush current. Also, it is not known at this time how local jurisdictions will interpret the NEC [2] regarding derating of conventional transformers in light of the new listings for harmonic-rated units.

9.17.2.3 Calculating Load Harmonics. Selection of a transformer derating factor or a K-factor rating requires characterization of the load current distortion. When measured harmonic data are not available for the load, it must be calculated. There is a tendency to calculate a higher than necessary K-factor when considering the individual load currents. Due to cancellation, the combined distortion of several loads is always less than the sum of individual loads. This reduction may be substantial when there is a large number and a diversity of nonlinear load types (see Fig 9-12). However, when loads are removed for whatever reason, the cancellation benefit produced by these loads is also removed. In many cases, this will not be a problem for a converatively load rated transformer, but can be a problem if the load rating is a marginal one.

Cancellation results when harmonics produced by different loads are phase shifted relative to each other. Impedance in branch circuit wiring, as well as isolation transformers or series inductors that may be incorporated in the loads, shift harmonic currents. A delta-wye connection in a transformer serving single-phase nonlinear loads will trap the balanced triplen harmonics and may substantially reduce distortion on the high side.

It is difficult to predict this harmonic diversity factor without modeling the nonlinear loads and the power distribution. Computer programs and methods that allow modeling and simulation are available. With more experience, these computer analysis tools are expected to provide diversity factors for typical loads in industrial and commercial power systems. If such factors are not

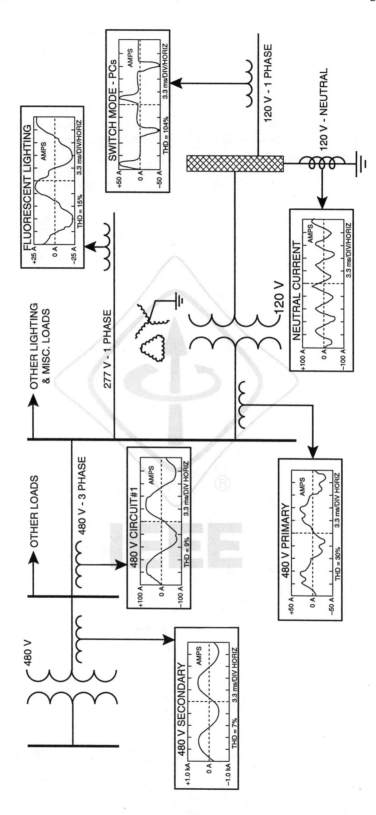

Fig 9-12
Example Distribution of Harmonic Levels in a Facility AC Distribution System

available, it is recommended practice to monitor the load current distortion and diversity relative to the load mix in a comparable facility.

In selecting a load K-factor, consider that K=1 indicates a linear load. Typical load K-factors for facilities containing large numbers of computers range from 4–9. K-factors for step-down transformers that serve almost exclusively nonlinear loads, such as personal computers, have been observed to range as high as 9–20.

9.17.3 Neutral Connection Derating for the Wye-Connected Transformer. Dry-type transformers used on three-phase applications (typically 208Y/120 Vac circuits), where a secondary neutral conductor is provided to a set of nonlinear (line-to-neutral) loads, should have the neutral path within them protected from the I^2R heating effects of the associated triplen harmonic currents which add in the neutral path. Transformers of this type should have 200% rated neutral connections.

9.17.4 Transformer Percent Impedance. Recommended practice is that the dry-type isolation transformers used with sensitive equipment have an impedance in the range of 3–5%, as calculated at the nominal line frequency. This impedance should not exceed 6% in any case. A *stiff* source (low impedance value) is an advantage in cases where loads are being served with high-peak current demand and large crest-factors, both of which are typical of sensitive electronic equipment. A lower value impedance also prevents *flat-topping* of the ac voltage waveform and reduces the overall problem of harCancellation results when harmonics produced by different loads are phase shifted relative to each other. Impedance in branch circuit wiring, as well as isolation transformers or series inductors that may be incorporated in the loads, shift harmonic currents. A delta-wye connection in a transformer serving single-phase nonlinear loads will trap the balanced triplen harmonics and may substantially reduce distortion on the high side.

It is difficult to predict this harmonic diversity factor without modeling the nonlinear loads and the power distribution. Computer programs and methods that allow modeling and simulation are available. With more experience, these computer analysis tools are expected to provide diversity factors for typical loads in industrial and commercial power systems. If such factors are not monic voltage waveform distortion on the transformer secondary.

9.17.5 Transformer Forward-Transfer Impedance. The forward transfer impedance of the IT should generally be selected to be as low as possible in the frequency range from the nominal power frequency up to about 5 kHz. This characteristic helps to reduce harmonic voltage waveform distortion on the output of the IT when it supplies nonlinear loads. Having too low forward-transfer impedance may not be beneficial as it may assist in the propagation of unwanted current, i.e., electrical noise, between the IT's windings.

9.17.6 Banked Transformers, Open Deltas, Tees, and Common Cores. Three-phase transformers should always be selected on the basis of having a common core (E-core) and should never be composed of banked single-phase transformers. Single-phase transformer cores quickly saturate and overheat when subjected to dc and triplen harmonic currents on their neutral path.

9.17.7 Making Conduit/Raceway Connections to Transformers. The typical dry-type transformer enclosures are designed and listed for the connection of conduits/raceways only at designated points. These points are generally below a given point in the enclosure where the ambient temperature has been tested and shown to not rise above the listed temperature range. Conduit/raceway connection to a transformer via holes cut into the top-cover, the rear or front-cover, or the ventilation screen in the bottom, are not recommended. When short sections of flexible metal conduit (FMC) are used between a transformer and non-flexible metal conduit/raceway, it is recommended that these sections be bonded across using low-inductance bonding means (they are shields and may carry HF currents).

9.17.8 Separation of Input from Output Wiring on Transformers. Because the typical IT is used for reduction of HF electrical noise, it is recommended that its primary (input) wiring not be routed in the same conduit/raceway as its secondary (output) wiring. It is also recommended that the input and output wiring inside the enclosure be separated as much as practical.

9.18 Wiring Devices. Wiring devices for sensitive electronic equipment are identified by the load equipment OEMs. They are generally called out on the associated OEM-provided installation data sheets.

9.18.1 NEMA, IEC, and Other Configurations. Several standards exist for which electrical connectors are configured and where connectors meeting the same general configuration (e.g., size, keying, and face pattern) are made by more than one OEM. In these cases, all are basically interchangeable even though mechanical construction and materials may differ significantly among the devices.

9.18.2 Wiring Device Conductor Terminations. Conductor terminations to wiring devices are a major source of problems due to careless assembly or improper assembly techniques. Unforeseen design and installation problems may create incompatibility between a conductor and its associated connector wiring terminal.

Most wiring termination problems can be controlled if the conductor and connector terminals are determined to be compatible with one another with regard to wire size range and aluminum/copper compatibility. Small receptacles often are provided with push-in wiring contacts in addition to screw-compression wiring contacts. Push-in contacts should not be used.

Three-phase receptacles capable of accepting neutral conductors of approximately 2× the cross-sectional area should be used where the neutral has been up-sized to compensate for harmonic currents from nonlinear loads. This recommendation prevents a possible localized failure due to I^2R overheating at the connectors.

9.18.3 Special Keying for Non-60 Hz Circuits. Circuits using connectors and operating at frequencies other than 60 Hz (United States standard) should not use connectors that are interchangeable with 60 Hz versions. Special keying is the recommended practice for dealing with this problem, as opposed to simply using a different configuration that may be considered unique at the given location. Such uniqueness often is not maintained over the lifetime of the site.

9.18.4 Grounding Configurations and Requirements. The receptacle and plug (cap) connected to an EGC shall have a dedicated and keyed pin reserved for the EGC conductor. It shall not be a connector where a non-EGC pin is field-assigned to the grounding function.

9.18.5 Neutral (Grounded Conductor) Configurations and Requirements. Only plugs and receptacles that are OEM-designed and configured to be connected to a neutral (i.e., grounded) conductor should be used for that purpose. Assigning the neutral conductor to some other pin on a connector is not recommended practice.

9.19 Pull and Junction Boxes. All pull and junction boxes should be metal (they are a part of the fault and shielding current path) and should not be equipped with concentric knockout forms for conduit connections. Enclosed boxes of the solid walled (unpunched) style are recommended. Consideration should be given to the environmental conditions in the selection of these boxes.

9.20 Metal Conduit and Metal Enclosed Wireway. All premises wiring feeder and branch circuit conductors serving sensitive electronic equipment should be fully enclosed by grounded metal. Each branch circuit should be in separate conduit. Solid wall metal conduit is recommended. Metal enclosed wireway also may be used.

9.20.1 Recommended Materials. Recommended conduit materials for most premises wiring purposes are rigid metal conduit (RMC), intermediate metal conduit (IMC), and electrical metallic tubing (EMT). These materials are listed in descending order of cost, conductivity, and shielding effectiveness. However, most sites do not experience the extreme conditions that make rigid metal conduit (RMC) a good choice to use. The typical site performs well with EMT.

Use of ferrous metal conduit is recommended (for shielding purposes) over non-ferrous conduits in all cases except for 380–480 Hz ac power circuits. Power of 400 Hz is best routed in nonferrous metal conduit for improved voltage-loss control because of ferrous conduits' higher losses at higher frequencies. Nonmetallic conduits and raceways are not recommended practice.

9.20.2 Metal Conduit for Signal Conductors. Solid wall metal conduit is recommended, particularly in the vicinity of power conductors. Signal conductors should not be run in the same conduit as power conductors. Conduits should be connected to building steel at multiple points along their length. When signal cables are run in wireways, the wireways should be fully enclosed.

9.20.3 Conduit Couplings. Coupling methods between sections of conduits should be properly installed to conduct noise currents with minimum voltage drop. Rigid and intermediate conduits should be threaded and related couplings should be made-up-tight during installation. Couplings utilizing a compression ring are recommended for non-threaded conduit. Clean connections are absolutely necessary. Set-screw type fittings are not recommended.

9.20.4 End-Terminating Fittings With Locknuts on Conduits. End-terminating fittings on conduit sections should conform to the same requirements as for couplings discussed above. In addition, end-fittings should be equipped with metal locknuts that are tightly locked-down against the surface of the terminating equipment's enclosure. For best results, set-screw equipped locknuts are recommended to ensure the continuance of low-level current paths for shielding purposes between the fitting and the terminating equipment's enclosure.

9.20.5 End-Terminating Fitting Use With Concentric Knockouts. Where any end-terminating fitting makes a mechanical connection to an equipment enclosure via a concentric knockout, it is recommended that the locknut assembly be equipped with a grounding-conductor connection means and a bonding jumper as well. The installation of a bonding jumper across the concentric ring tabs (which are not a reliable means of providing a good grounding or bonding path for HF currents and safety currents) is recommended. The jumper should be terminated directly to the equipment grounding bus-bar in the equipment, or if a bus-bar is not available, directly to the metal enclosure using a properly prepared grounding-surface and joint.

"Daisy-chain" busing of a bonding jumper between more than one grounding or bonding locknut is not recommended, but if it is used then the jumper should be bonded to the equipment grounding bus-bar at each end. The bonding jumper should not be snaked back and forth in an attempt to use one jumper for all connections. A dedicated bonding jumper should be used for each row of connections.

9.20.6 Special Requirements for Liquid-tight Terminating Fittings. In all cases where a liquid-tight form of termination is employed between a conduit and an equipment enclosure, the use of the associated listed sealing ring or gland assembly for the interface between the fitting and enclosure is recommended. In some cases, this ring or gland is not only the sealing method but is also an integral part of the grounding path.

9.20.7 Use of Terminating Fittings With Reducing Washers. Conduits are frequently required to be terminated to equipment that have knockout openings that are too large for the intended conduit. The use of reducing washers on sensitive electronic equipment applications is not recommended. Their usage is a problem at high frequencies and with low currents. It is a critical problem where the washer set is applied over painted or non-conductively plated metal surface. If a fitting or reducing washer is used, recommended practice is to bond the fitting to the enclosure with a separate, low-inductance jumper.

9.21 References. This standard shall be used in conjunction with the following publications. When the following standards are superseded by an approved revision, the revision shall apply:

[1] ANSI/EIA/TIA 571-1991, Environmental Considerations for Telephone Terminals.[27]

[2] ANSI/NFPA 70-1993, National Electrical Code.[28]

[3] ANSI/NFPA 75-1992, Protection of Electronic Computer/Data Processing Equipment.

[4] ANSI/NFPA 780-1992, Lightning Protection Code.

[5] ANSI/UL 96-1985, Lightning Protection Components.[29]

[6] IEEE Std C57.12.91-1979, IEEE Test Code for Dry-Type Distribution and Power Transformers.[30]

[7] IEEE Std C57.110-1986, IEEE Recommended Practice for Establishing Transformer Capability When Supplying Nonsinusoidal Load Currents (ANSI).

[8] IEEE Std C62.36-1991, IEEE Standard Test Methods for Surge Protectors Used in Low-Voltage Data, Communications, and Signaling Circuits.

[9] IEEE Std C62.41-1991, IEEE Recommended Practice on Surge Voltages in Low-Voltage AC Power Circuits (ANSI).

[10] IEEE Std C62.45-1987, IEEE Guide on Surge Testing for Equipment Connected to Low-Voltage AC Power Circuits (ANSI).

[11] IEEE Std 141-1986, IEEE Recommended Practice for Electric Power Distribution for Industrial Plants (Red Book) (ANSI).

---

[27] EIA/TIA publications are available from Global Engineering, 1990 M Street NW, Suite 400, Washington, DC, 20036, USA.

[28] NFPA publications are available from Publications Sales, National Fire Protection Association, 1 Batterymarch Park, P.O. Box 9101, Quincy, MA 02269-9101, USA.

[29] UL publications are available from Underwriters Laboratories, Inc., 333 Pfingsten Road, Northbrook, IL 60062-2096, USA.

[30] IEEE publications are available from the Institute of Electrical and Electronics Engineers, Service Center, 445 Hoes Lane, P.O. Box 1331, Piscataway, NJ 08855-1331, USA.

[12] IEEE Std 142-1982, IEEE Recommended Practice for Grounding of Industrial and Commercial Power Systems (Green Book) (ANSI).

[13] IEEE Std 399-1990, IEEE Recommended Practice for Industrial and Commercial Power Systems Analysis (Brown Book) (ANSI).

[14] IEEE Std 446-1987, IEEE Recommended Practice for Emergency and Standby Power Systems for Industrial and Commercial Applications (Orange Book) (ANSI).

[15] IEEE Std 519-1992, IEEE Guide for Harmonic Control and Reactive Compensation of Static Power Converters (ANSI).

[16] UL 96A-1982, Installation Requirements for Lightning Protection Systems.

[17] UL 497-1991, Protectors for Paired Conductor Communications Circuits.

[18] UL 497A-1990, Secondary Protectors for Communications Circuits.

[19] UL 1449-1985, Transient Voltage Surge Suppressors.

[20] UL 1950-1989, Information Technology Equipment Including Electrical Business Equipment.

[21] Federal Information Processing Standards Publication 94: *Guideline on Electrical Power for ADP Installations*, Sept. 21, 1983.[31]

9.22 Bibliography

[B1] Lai, J. S. and F. D. Martzloff, "Coordinating Cascaded Surge-Protective Devices," *Proceedings*, IEEE/IAS Annual Meeting, October 1991, pp. 1812–19.

[B2] Martzloff, F. D., "Coupling, Propagation, and Side Effects of Surges in an Industrial Building Wiring System," *IEEE Transactions on Industry Applications*, Vol. IA-26, No. 2, March/April 1990, pp. 193–203.

[B3] Morrison and Lewis, *Grounding and Shielding for Facilities*, J. Wiley & Sons, 1990.

[B4] Zavadil, R., M. F. McGranaghan, G. Hensley, and K. Johnson, "Analysis of Harmonic Distortion Levels in Commercials Buildings," *Proceedings, PQA '91, First International Conference on Power Quality: End-Use Applications and Perspectives.*

---

[31] FIPS documents are available from the National Technical Information Service (NTIS), U. S. Dept. of Commerce, 5285 Port Royal Rd., Springfield, VA 22161.

Appendix
Interpreting and Applying
Existing Power Quality Studies

(This appendix is not part of IEEE Std 1100-1992, but is included for information only.)

A1. Introduction

Power line monitoring has been advocated as a method to determine power quality at an existing or proposed computer site. The results of the study are generally used as a basis to determine what, if any, power conditioning equipment is required. While individual site monitoring is a logical first step, it can also be costly and time-consuming. Since severe power disturbances may occur infrequently or on a seasonal basis, monitoring periods of six months to a year may be required to provide an accurate power disturbance profile.

An assumption made in using individual site monitoring as a basis for selecting power conditioning equipment is that the recorded disturbances are representative of future occurrences. Any changes made to the site, neighboring sites, or even the power utility system can alter the disturbance profile.

A2. Analyzing Existing Studies

Comprehensive power studies conducted at a number of sites across the country provide a more generalized view of power disturbances. The results of these studies can be used in conjunction with individual site monitoring to determine the types of disturbances likely to threaten sensitive electronic equipment. Studies of United States power quality include the Allen-Segall study [B1],[32] the Goldstein-Speranza study [B2], and the T. S. Key study [B3]. The Allen-Segall and Goldstein-Speranza studies have been recognized as two of the most comprehensive studies. A cursory comparison of the results (Table A1) would lead one to the conclusion that a significant change in power disturbances at computer sites has occurred between 1972 (end of Allen-Segall study) and 1979 (end of Goldstein-Speranza study).

[B1] reported 88.3% of the recorded disturbances to be surges, 11.2% sags, and 0.47% interruptions; [B2] reported 87% of the disturbances as sags, 7.4% surges, 0.7% swells, and 4.7% interruptions. A look at the monitoring thresholds used in each study (Table A2) helps to explain why the number of surges appears to have decreased and the number of sags increased. [B2] used a sag threshold of −4% while [B1] used a threshold of −10%, which would explain the higher percentage of sags in [B2]. Sags between −4% and −10% would

---

[32] The numbers in brackets correspond to those in the bibliography in A5.

not be reported in [B1]. It should also be noted that the monitor used in [B2] uses a moving rms voltage reference to determine sag and swell events. A 5% change in the rms voltage would therefore be recorded as a sag even though the voltage remained within acceptable operating limits. This point should be kept in mind when using this data to recommend voltage regulation equipment.

Table A1
Comparison of National Power Quality Studies

	Goldstein-Speranza [B2]	Allen-Segall [B1]
Date of study	1977–1979	1969–1972
Monitor months	270	147
Number of sites	24*	29
Oscillatory, decaying surges	**	48.8%
Surges	7.4%	39.5%
Sags	87.0%	11.2%
Swells	0.7%	0.0%
Interruptions	4.7%	0.47%

* Refers to sites confined to Bell System.
**Included as surges.

Table A2
Monitor Threshold Settings

	Goldstein-Speranza [B2]	Allen-Segall [B1]
Oscillatory, decaying transients (surges)	200 V	±15%
Surges	200 V	±10%
Sags/Swells	±5 V	±10%

The linear power supplies, used in electronic equipment at the time these studies were conducted, were more sensitive to changes in the ac voltage than are the switch-mode power supplies used in most electronic equipment today.

Most modern electronic equipment today can tolerate as much as a 10% variation in the ac supply voltage without problems. The threshold for surges used by [B2] (118%) was much greater than the threshold used by [B1] (10%). Therefore, one would expect a significant reduction in the percentage of surges reported by [B2] as compared to the [B1] study.

APPENDIX

The increase in the percentage of interruptions reported by [B2] may be explained by the shift in the total disturbances observed due to the other threshold changes. Percentages can be a very misleading basis for comparison unless all conditions are equal. For example, the incidence rate of interruptions in both studies is very similar, even though the percentages were an order of magnitude apart.

Both studies presented summaries and statistical analysis of their data in different ways. [B1] used only the observed disturbances as a data base and presented results in various incidence rate plots. Incidence rates of sags and surges at different thresholds are given by [B1], which allow a more direct comparison with the disturbance thresholds of [B2].

[B2] presented a statistical model of the disturbance rates to predict the incident rates at pre-defined thresholds. The thresholds of the [B2] model were different from the monitoring thresholds and were chosen at levels generally expected to cause computer system problems: surges greater than 200 V, sags greater than −20% of nominal, and swells greater than +10% of nominal. The [B2] model states the disturbance rates in probabilistic terms, such as 50% of the sites will have less than "x" disturbances per year, or 90% of the sites will have less than "y" disturbances per year. When the disturbance rates at the same thresholds are compared for the [B1] data and the [B2] model, the results are surprisingly similar (Table A3).

Table A3
Normal Mode Power Disturbances per Year

	Goldstein-Speranza [B2]	Allen-Segall [B1]	Combined Data
Surges (> 100% of peak)	8 (15%)	12 (27%)	10 (21%)
Sags (> −20% rms)	36 (68%)	25 (57%)	30 (62%)
Swells (> +10% rms)	2 (4%)	0 (0%)	1 (2%)
Interruptions (one phase or more)	7 (13%)	7 (16%)	7 (14%)
TOTAL	53 (100%)	44 (100%)	48 (100%)

Common-mode power disturbances were not included in either [B1] or [B2]. At the time these studies were conducted, the linear power supplies used in equipment provided some level of protection against these disturbances. The switch-mode power supplies used today tend to be more susceptible to common-mode disturbances. Disruptive common-mode disturbances can be expected to occur at least as often as severe normal-mode surges. If common-mode disturbances are estimated to occur at twenty-four times per year, Table A4 represents the expected power disturbances per year that may disrupt computer system operations at a typical site within the United States.

Table A4
Expected Disruptive Power Disturbances per Year

Surges (> 100% of peak)	10 (14%)
Sags (> −20% rms)	30 (42%)
Swells (> +10% rms)	1 (1%)
Interruptions (one phase or more)	7 (10%)
Common-mode disturbances	24 (33%)
TOTAL	72 (100%)

A3. Effects of Power Disturbances on Sensitive Electronic Systems

The T. S. Key study and test results [B3] have shown that the susceptibility of sensitive electronic equipment to power disturbances can be divided into three areas:
(1) Momentary overvoltages
(2) Sags and momentary interruptions
(3) Interruptions

A3.1 Susceptibility to Momentary Overvoltages. Momentary overvoltages and high-frequency electrical noise may have an indirect effect on equipment performance depending on the operating conditions and severity of the disturbance; however, a recorded disturbance may or may not coincide with the time of a computer system malfunction. If there is an effect, it may take the form of a processing "stop" which can be the failure of an individual piece of equipment. To resume processing, an automatic "retry," operator "reset," or in the worst case, field engineer "repair" is required. Typical examples of "stops" are parity errors, keyboard lockups, read/write errors, and lost file labels. Many of these malfunctions can appear to be hardware or software problems instead of power quality problems. These types of disturbances can also have long-term effects on equipment performance. Depending on the severity of the disturbance, component degradation can occur and equipment malfunctions may not start until several months after installation.

A3.2 Susceptibility to Sags and Interruptions. Sags and interruptions normally have a direct and predictable effect on sensitive electronic equipment. Built-in sag (and overvoltage) sensors are pre-set to power down the equipment if the voltage exceed the limits. The combined effect of the magnitude and duration of voltage excursions determines whether or not the electronic equipment powers down (crashes).

Some newer equipment will sense the loss of input ac power, initiate an orderly shutdown, and automatically power on when the ac source is restored. In a study of 100 computer failures (crashes), [B3] used power-line monitoring, system operation logs, National Weather Service data, and local utility data to determine the cause of the power related failure. The results are shown in Table A5. These results confirm that weather is a major cause of disruptive sags and interruptions.

Sags were found to disrupt computer operations four times as often as interruptions, which closely agrees with the results of the combined [B1] and [B2] studies in which sags outnumbered interruptions 4.3 to 1.

Table A5
Causes of 100 Disruptive Undervoltages

Cause	Sags	Interruptions	Total
Weather (Wind and lightning)	37	14	51
Utility equipment	8	0	8
Construction or traffic accident	8	2	10
Animals	5	1	6
Tree limbs	1	1	2
Unknown	21	2	23
TOTAL	80	20	100

A4. Power Disturbances by Type of Power System

The [B3] study classified disruptive sags and interruptions by the type of utility power system. The results are shown in Table A6. A similar analysis was performed by [B1], which yielded essentially the same results.

Both studies found a significant advantage in underground distribution systems and a significant disadvantage in overhead "spot" networks (where exposure to system faults are higher). As expected, networks have a significantly lower incidence of interruptions.

Table A6
Disruptive Undervoltages per Year

Type of Power System	Sags*	Interruptions*	Total*
Overhead radial	12	6	18
Overhead spot network	22	1	23
Combined overhead	16	4	20
Underground radial	6	4	10
Underground network	5	0	5
Combined underground	5	2	7
Total combined	11	3	14

*Weighted average based on months monitored.

A5. Bibliography

[B1] Allen, G. W. and D. Segall, "Monitoring of Computer Installations for Power Line Disturbances," *IEEE Winter Power Meeting Conference Paper*, WINPWR C74 199-6, 1974 (abstract in *IEEE Transactions on PAS*, Vol. PAS-93, Jul/Aug 1974, p. 1023).

[B2] Goldstein, M. and P. D. Speranza, "The Quality of U. S. Commercial ac Power," *INTELEC (IEEE International Telecommunications Energy Conference)*, 1982, pp. 28–33 [CH1818-4].

[B3] Key, T. S. "Diagnosing Power Quality Related Computer Problems," *IEEE Transactions on Industry Applications*, Vol. IA-15, No. 4, July/Aug 1979.

Index

A

Abbreviations, 35-36
AC input failure and return test (UPS), 194
Adjustable frequency drive (AFD) motors, 145-147
Air conditioning requirements, power system, 179
ALVRT (Automatic line voltage regulating transformer), 231
American National Standards Institute/National Fire Protection Association. *See* ANSI/NFPA standards
Ammeters
 definition of, 26
 clamp-on, 104-105, 129
 current transformer (CT), 107, 109
 direct-reading, 26, 107
 Hall effect, 108-109
 true RMS, 107, 109
Ampacity, 228-230
Amplifiers, 88
Amplitude variations, 42
Analog integrated circuits, 89
ANSI/NFPA (American National Standards Institute/National Fire Protection Association) standards, 103, 108, 116, 127, 201, 206, 230
Arcing, 78, 83
ASAI (Average Service Availability Index), 68
Assembly-line restart times, excessive, 146
Atmospheric charge redistribution lightning. *See* Lightning protection
nonarcing, 82, 145-146
Automated cash register and checkout case history, 155-156
Automated forward and reverse transfer (UPS), 189
Automated office, case history, 150-151
Automatic line voltage regulating transformer (ALVRT), 231
Avalanche diodes, 160, 165
Average Service Availability Index (ASAI), 68

B

Batteries, 173-177, 179-180, 185, 194-196
Blackout (term to avoid), 34
Bonding, 104-105, 205-207, 209, 226
 concepts, 95-96
 definition of, 25
 lightning protection subsystem, 93-94
 neutral-ground, 127
 signal reference grid, 213, 215
 straps, 215-216

Bond reactance, 96
Branch circuits, 204
Brownout (term to avoid), 34
Buck-boost regulators, 167-168
Building AC distribution system impedance, 61
Building structural steel, 205-206, 213, 215, 217
Buried cables, 86-87, 245-246
Buried ring-ground, 206, 229-230
Bus bars, 156, 217, 229, 238
Busways, 204

C

Cable accessed television (CATV), 217
Cables, buried, 86-87
Calculated reliability (MTBF), 183-184
Capacitance, lumped, 57, 66
Capacitive coupling, 86
Capacitive reactance, 63-64
Capacitors, 59, 71, 146-147, 151
Case histories, 143-156
 electrical inspection problems, 155-156
 electronic loads, 149-150
 equipment misapplication, 156-157
 life-safety system problems, 156
 premises switching generated surges, 147-148
 premises-wiring related problems, 151-154
 radiated EMI problems, 155
 transient voltage surge suppression network design, 154-155
 utility-sourced power quality problems, 145-147
CATV (cable accessed television), 217
Central office trunk cable (COTC), 217
Chemical processing plant case histories, 153-154
Circuit breakers, 43, 103, 172, 186, 204, 228
Clamp-on ammeters, 104-105, 129
Clamp-on current transducers, 112
Clean ground (term to avoid), 34
Clean power (term to avoid), 34
Codes and standards, 21, 51-54. *See also* IEEE
Color-coding circuits, 210
Commercial power, 25
Common-mode noise, 25, 162
Common-mode transient, 162
Communication systems, 216-217, 220-222
Component wear and aging, 195-196
Computed tomography (CT) scan case history, 147
Computer-based harmonic analysis, 120
Computer grade ground (term to avoid), 34
Computer power centers, 172
Computers, 40, 46, 49-51, 150-152, 155
 CAD/CAM graphics system case history, 153

data error case history, 150
distributed throughout a facility, 46
mainframe computer rooms, 46, 149, 179, 186, 201-202
personal computers, 36, 152
process control case histories, 152-153
voltage tolerance envelope, 47
Concrete encased ground electrode, 27
Conditioning, power. See Power line conditioners
Conducting barrier (term to avoid), 34
Conductors
  arcing distances, 83
  buried, 86-87
  measurement, 104-105, 127-128
  overhead, 87, 245-246
  self-resonance, 66-68
Conduit, 105, 206-207, 209, 212, 226, 236-239
Constant voltage transformer (CVT), 168-169. See also Ferroresonant transformers
Costs, 22, 141, 159
  installation, 185-186
  operation, 41, 186-187
COTC (central office trunk cable), 217
Counterpoise ground (term to avoid), 34
Coupling
  definition of, 26
  far field, 86
  free space, 84
  mechanisms, 84-86
Crest factor, 26, 70, 109, 178
Critical load, 26
CT (computed tomography) case history, 147
CT (current transformer) ammeters, 107-109
Current
  distortion, 70, 181
  eddy, 74-75, 77
  fault, 71-72
  inrush, 29, 71-72, 107, 180
  measurement, 104-105, 109
  ripple, 180
  transducers, 112
Current transformer (CT) ammeters, 107-109
CVT. See Constant voltage transformer

D

D'Arsonval meters, 107
Data disruption, 72, 87-88, 150
Data lines, 46, 70, 153, 215-217, 220-222
Data port, 50
DC current, 70, 109
Dedicated feeders, 129
Dedicated ground (term to avoid), 34
Definitions, 25-36
  abbreviations, 35-36
  terms to avoid, 34-35
Degradation failure, 26
Department of Defense, 51
Derating, component, 74-75, 228, 232
Design and installation practices, recommended, 21-22, 199-239

automatic line voltage regulating transformer, (ALVRT), 231
branch circuits, 204
computer room wiring and grounding, 201-202
conduit and wireway, 237-239
conflicting philosophies, 21-22, 25, 34-35, 39, 41-43, 45-46
dedicated and shared circuits, 202
dry-type transformers, 231-236
feeders, 129, 202-204
400 Hz power systems. See 400 Hz power systems
general discussion, 199-201
grounding. See Grounding
harmonic current control. See Harmonics
lightning surge protection. See Lightning protection
panelboards, 229-231
power distribution units (PDUs) for AC power-load interface, 231
power factor improvement, 204-205. See also Power factor
pull and junction boxes, 237
single-phase input conditions on three-phase load equipment, avoiding, 204
switchboards, 229
transfer switching, specialized AC source, 205
wiring devices, 236. See also Wiring
Differential-mode noise. See Transverse-mode noise
Digital integrated circuits, 89
Diodes, 89, 160, 165
Dip (term to avoid), 34
Direct-reading ammeters, 26, 107
Dirty ground (term to avoid), 34
Dirty power (term to avoid), 34
Displacement power factor, 31, 73
Dissymmetry, 42
Distortion, current. See Voltage distortion
Distortion, voltage. See Voltage distortion
Distortion factor, 26
Disturbances, power. See Power disturbances
Documentation, 130-135, 215
Drawings, mechanical and electrical, 205
Dropout, 26
Dropout voltage, 26
Dry-type transformers, 231-236
Dynamic response, 188-189

E

Earth electrode grounding subsystem, 90, 209, 216
Earth ground resistance testers, 104-105, 110-111
Eddy current, 74-75, 77
Efficiency (of a power system), 26, 186-187, 194
Electrical fast transient (EFT), 61, 89
Electrical inspection problem case histories, 155-156
Electrical safety listing avoidance, 192

Electrodynamometers, 107
Electromagnetic interference. *See* EMI
Electromagnetic shielding, 95
Electronic equipment. *See* Sensitive electronic equipment
Electronic loads. *See* Load; Load and power source interactions
Electrostatic discharge (ESD), 45, 70, 82-84, 122, 140-141
Electrostatic shielding, 94, 161
EMI (electromagnetic interference), 44, 49, 51-52, 122, 140, 205
  case history, 155
Enclosed EMI/EMC areas, 51-52
Energy effects, 48
Engineering laboratory case history, 150-151
Enhancement, power. *See* Equipment specification and selection
Environmentally induced surges, 81-89
  electrostatic discharge, 82-84
  lightning induced, 81-82
  nonarcing atmospheric charge redistribution, 82
Equipment, power enhancement. *See* Equipment specification and selection
Equipment, protective, 103, 108, 116, 127
Equipment, sensitive electronic. *See* Sensitive electronic equipment
Equipment grounding conductor, 26
  isolated, 29
Equipment safety grounding conductor (term to avoid), 35
Equipment specification and selection, 159-197
  bonding. *See* Bonding
  case histories. *See* Case histories
  commonly used devices, 160-177
  computer power centers, 172
  conflicting information and perspectives concerning, 21-22, 25, 34-35, 39, 41-43, 45-46
  cooperation between different parties, need for, 22, 39-40, 57, 124-125
  current harmonic filters, 160, 162, 164-165
  design practices. *See* Design and installation practices, recommended
  400 Hz power systems. *See* 400 Hz power systems
  general considerations, 21-24, 39-46, 159
  grounding. *See* Grounding
  impedance considerations. *See* Impedance
  installation practices. *See* Design and installation practices, recommended
  isolation transformers, 154, 160-164, 172, 207-208
  load and power source interactions. *See* Load and power source interactions
  maintenance. *See* Maintenance
  noise filters, 160, 162, 164
  power line conditioners. *See* Power line conditioners
  procurement specifications. *See* Procurement specifications
  safety considerations. *See* Safety considerations
  site surveys and power analyses. *See* Power quality studies, existing; Site surveys and site power analyses
  specification writing, 190-193
  standards, 21, 51-54. *See also* IEEE
  standby power systems, 161-163, 172-174
  surge suppressors, 43, 160, 162, 165-166, 205
  transformers. *See* Transformers
  uninterruptible power supply. *See* Uninterruptible power supply
  verification testing. *See* Verification testing
  voltage regulators. *See* Voltage regulators
  see also Sensitive electronic equipment
Equipotential plane, 92
ESD. *See* Electrostatic discharge
Event indicators, 113
Expansion joints, 207
Expert systems, 120-122

F

Facility planner's considerations, 178
Factor, power. *See* Power factor
Failure, degradation, 26
Failure mode, 26
Faraday cage, 52
Faraday shield, 161
Far-field coupling, 86
Fault currents, 71-72
Federal Information Processing Standards publications, 53
Feeders, 129, 202-204
Ferroresonant transformers, 168-169, 173, 177
Fiber optic communication links, 153, 223
Field reliability data, 184
Flashover effects, 48
Flat strip SRG, 212-213
Flexible metal conduit (FMC), 236
Flicker, 27
Floating voltage, 180
Flooring, raised, 212-213
FMC (flexible metal conduit), 236
Form factor, 27
Forward-transfer
  automatic (UPS), 189
  impedance, 27, 59
400 Hz power systems
  component derating, 228-229
  conductor ampacity, 228
  grounding and shielding, 226
  location of, recommended, 226
  wiring losses, controlling, 226, 228
Frame ground (term to avoid), 35
Free-space coupling
  capacitive, 86
  inductive, 84-86
Frequency
  deviations, 27, 42, 70, 72, 162
  measurement, 105

performance range, 57-58
power/safety range, 57
surge, 88-89, 92-93
Frequency shift (term to avoid), 35
Fuses, 204, 228

G

Glitch (term to avoid), 35
Ground, definition of, 27
Ground, high-frequency reference. *See* Signal reference structure
Ground, radial, 28
Ground, ufer. *See* Ground electrode, concrete encased
Ground conductors, 152, 155-156
Ground electrode, 27
  concrete encased, 27
Ground grid, 27
Ground impedance testers, 27, 104, 109-110, 128
Grounding, 22, 46, 154, 205-218, 237
  AC system, 207, 217
  annotating mechanical and electrical drawings, 205
  bonding, 93-94, 104-105, 206-207, 209, 213, 215
  building structural steel, 205-206, 213-215, 217
  buried ring-ground electrode system, 206, 229-230
  communication systems, 216-217
  conduits and raceways, 206-207, 209
  discontinuity, 152-154
  earth electrodes, 90, 209, 216
  expansion joints, 207
  400 Hz power systems, 226
  galvanized construction channel as bus bar, 217
  high-frequency ground referencing systems, 210, 212-214
  isolated/insulated ground (IG) method, 207, 209
  lightning protection. *See* Lightning protection
  measurement instruments, 103
  mechanical equipment in electronic equipment areas, 206
  panelboard, 230
  safety, 22-23, 90, 205
  separately derived AC sources, 207
  solidly grounded AC supply systems, 205, 207
  SRG (signal reference grid), 210, 212-216
  SRS (signal reference structure), 91-93, 210, 212, 216
  survey, 126-135
  topological, 52
  uninsulated conductors, 218
  UPS, 217
Ground loop, 28, 154, 161, 214
Ground resistance testers, 104-105, 110-111
Ground window, 28

H

Hall-effect ammeters, 108
Hardware destruction, 88
Hardware stress, 88
Harmonics, 40, 59, 65, 73-74, 172, 181, 204, 229
  analysis, computer-based, 120-121
  calculating, 233-235
  computer-based analysis, 120
  definitions of, 26, 28
  filters, 160, 162, 164-165
  measurement, 105
  test (UPS), 195
  total harmonic distortion, 26, 188
  triplen, 73, 76, 150, 230
Heating (overheating), 76-77, 106
High-frequency ground referencing systems, 210, 212-214
High-speed transistors and integrated circuits, 89
Historical perspective, 21-22, 39-40
Humidity measurements, 122, 140

I

IEC (International Electrotechnical Commission), 52-54, 236
IEEE (Institute of Electrical and Electronics Engineers)
  Color Book series, 23, 46
  standards, 103, 108, 116, 127, 205, 219, 232
  Working Groups, 24, 45
IG (isolated/insulated ground) method, 207, 209
Impedance, 57-68, 152, 235
  building AC distribution system, 61
  earth grounding electrode, 216
  equipment grounding conductor, 128
  forward transfer, 27, 59
  400 Hz power systems, 227
  frequencies of interest, 57-58
  load, 61-63
  measurement, 104-105
  neutral conductor, 128
  output, 31, 59-61
  power source, 58-61
  resonance considerations, 64-68
  reverse transfer, 31
  testers, 27, 104, 109-110, 128
Impulse. *See* Transient
Inductance, lumped, 57, 66
Inductive coupling, 84-86
Inductive reactance, 63
Information technology systems. *See* Computers
Infrared detectors, 106
Input power factor, 28, 180-181
Input soft-start, 180
Input transient suppression, 187
Input voltage range, 28, 188
Inrush current, 29, 71-72, 109, 180
Inspection, visual, 193

INDEX

Installation. *See* Costs, installation; Design and installation practices, recommended
Institute of Electrical and Electronics Engineers. *See* IEEE
Instrumentation, 103-122
   ammeters. *See* Ammeters
   computer-based harmonic analysis, 120
   current measurement considerations, 104-105, 109-110
   earth ground resistance testers, 104-105, 110-111
   electromagnetic interference (EMI) probes, 122
   electrostatic discharge meters, 122
   expert systems, 120-122
   ground circuit impedance testers, 27, 104, 109-110, 128
   infrared detectors, 106
   oscilloscope measurements, 104-105, 111-112, 136-139, 141-142
   power disturbance monitors, 104-105, 112-120
   radio frequency interference (RFI) probes, 70, 122
   receptacle circuit testers, 32, 109
   spectrum analyzers, 104-105, 120
   temperature and relative humidity measurement, 122, 140, 156-157
   voltmeters, 104-107
   wiring and grounding measurement instruments, 103
Integrated circuits, 89
Internal impedance, 58-59
International Electrotechnical Commission (IEC), 52-54, 236
International standards, 53-54. *See also* Standards
Interruptions, 29, 105, 162, 241-246
Inverters, 174-175, 190
Isolated equipment ground, 29
Isolated/insulated ground (IG) method, 207, 209
Isolated redundant systems, 182
Isolation, 29
Isolation transformers, 154, 160-164, 172, 207-208

J

Junction boxes, 237

K

K-factor rated transformers, 75-76, 233
Knowledge-based software, 120

L

Laplace equation, 94
LC line filters, 71, 164, 207, 226
LDC (Line drop compensator), 226, 228
Levels 1 to 3 site surveys, 125-126
Liability, 192
Lightning protection, 22, 43-44, 81-84, 86-87, 93-94, 142, 145-146, 152, 215-223
   data cabling and equipment, 220-221
   premise, 219
   service entrance, 218-219
   telecommunication system, 221-225
   UPS, 220
Linear load, 29
Line de-coupler, 110-111
Line drop compensator (LDC), 226, 228
Load
   critical, 26
   impedance, 61-63
   isolation, 187
   linear, 29
   nonlinear, 30, 45, 73, 164
   rating, 178
   step, 71
   switching, 44-45
   tests, 194-195
   three phase versus single phase, 200, 204
   tolerance, 70
   unbalanced, 34, 188
Load and power source interactions, 68-77
   case studies, 149-151
   corrective means, 77
   separating dissimilar loads, 202-203
   steady state voltage distortions, 73-77
   tolerances, 70, 89
   transient voltage disturbances, 71-72
Long-term power monitoring, 141-142
Low-noise transistors and diodes, 89
Low-power transistors and signal diodes, 89
Lumped capacitance, 57, 66
Lumped inductance, 57, 66
Lumped resistance, 57

M

Magnetic synthesizers, 169-170, 173
Maintenance, 21
   costs, 187
   long-term, 192
   preventive, 195
   restoring system operation after failure, 196-197
   wear and aging of components, 195-196
Maintenance-free batteries, 179, 185
Manufacturer's experience with power equipment, 184-185
Manufacturing plant case histories, 147, 150
Materials, specification and selection of. *See* Equipment specification and selection
MCT (metal cable tray), 212
Mean time between failures (MTBF), 183-184
Mechanical equipment in electronic equipment areas, grounding, 206
Medical clinic case history, 147
Medium power transistors, 89
Metal cable tray (MCT), 212
Metal conduit. *See* Conduit

M-Gs (motor generators), 161-162, 170-175, 190, 196
Motor generators (M-Gs), 161-162, 170-175, 190, 196
Motors, 61, 71-72
MTBF (mean time between failures), 183-184
Multimeters, 104-105, 108

N

National Bureau of Standards, 53
National Electrical Code (NEC), 21, 36, 41, 51-54, 126-127, 129, 136, 152, 155-156, 199-201, 205-206
   violations of, 21
National Electrical Manufacturers Association (NEMA), 53, 236
National Fire Protection Association (NFPA) standards, 103, 108, 116, 127, 201, 206, 230
National Institute of Standards and Technology (NIST), 53
Natural electrode (term to avoid), 35
NEC. *See* National Electrical Code
NEMA (National Electrical Manufacturers Association), 53, 236
NEMP (nuclear electromagnetic pulse), 45, 84
Neutral conductor sizing, 127-128
Neutral-ground bond, 127
NFPA (National Fire Protection Association) standards, 103, 108, 116, 127, 201, 206, 230
Noise (audible), equipment, 179
Noise (electrical), 21-22, 50-51
   common mode, 25, 162
   definition of, 30
   filters, 160, 162, 164
   measurement, 104-105
   normal mode, 162
   protection, 48-49
   transverse mode, 34, 162
Nonarcing atmospheric charge redistribution, 82, 145-146
Nonlinear loads, 45, 73, 164
   current, 30
   definition of, 30
Normal mode noise, 162
Normal mode transient, 162
Notches, 30, 105, 162
   case histories involving, 148-149
Nuclear electromagnetic pulse (NEMP), 45, 84

O

Occupational Safety and Health Administration (OSHA), 36, 103, 108, 116, 127
Off-line power systems, 161-163, 172-174
Operational specifications, 187
Operation cost considerations, 186-187
Oscilloscope measurements, 104-105, 111
OSHA (Occupational Safety and Health Administration), 36, 103, 108, 116, 127
Outage. *See* Interruption
Output impedance, 31, 59-61
Output voltage distortion, 188
Output voltage regulation, 188
Overcurrent protection, 22, 46, 151
Overhead conductors, 87, 245-246
Overload capability test, 195
Overload capacity and duration, 187-188
Overvoltage, 70, 105, 162, 244
   definition of, 31

P

Panelboards, 229-231
Parallel resonance, 65-66
Parallel systems, 181-182
PC (personal computer), 36, 152. *See also* Computers
PDU (power distribuition unit), 36, 231
Performance range, 57-58
Periodic function, 27
Personal computer (PC), 36, 152. *See also* Computers
Personnel hazards, 46, 50, 205. *See also* Safety
Phase imbalance, 70
Phase shift, 72, 77, 178
   definition of, 31
Ports, 50, 138
Power analysis. *See* Power quality studies, existing; Site surveys and site power analyses
Power conditioning equipment. *See* Power line conditioners
Power diodes, 89, 160, 165
Power distribution unit (PDU), 36, 231
Power disturbances, 244-246
   definition of, 31
   monitors, 31, 104-105, 112-113, 136-139, 141-142
   protection against, 47-50. *See also* Noise protection; Transient voltage disturbances
Power enhancement equipment. *See* Equipment specification and selection
Power factor, 70, 78, 146-147, 151, 178
   correction, 39, 204-205
   displacement, 31, 73
   input, 28, 180-181
   total, 31, 77
Power line conditioners, 123, 136-137, 139, 142, 146, 166, 169-172, 200, 241
   magnetic synthesizers, 161, 169-170, 173
   motor generators, 161-162, 170-175, 190, 196
Power quality, 21, 159
   concept, 40-42
   definition of, 31
   disturbances, case histories, 145-147
   disturbances, classification of, 42, 241
   disturbances, origin of, 42-45
   site surveys. *See* Site surveys and site power analyses
Power quality studies, existing, 241-246

analyzing, 241-243
effects of power disturbances on electronic equipment, 244
power disturbance by type of power system, 245-246
Power/safety range, 57
Power SCRs, 89
Power source impedance, 58-61
  internal, 58-59
  forward transfer, 27, 59
  output, 31, 59-61
Power supply port, 50
Preinsertion inductor, 146-147
Preventive maintenance. *See* Maintenance
Procurement specifications, 177-190
  facility planner's considerations, 177-181
  installation cost considerations, 185-186
  operation cost considerations, 186-187
  power technology considerations, 189
  reliability considerations, 181-185
  specifying engineer's considerations, 187-189
  transfer characteristics, 189
  writing, 190-193
  see also Equipment specification and selection
Protective equipment, 103, 108, 116, 127
Pull boxes, 237
PWM inverters, 190

Q

Quality, power. *See* Power quality

R

Raceway, 105, 206-207, 209, 212, 226, 236-239
Radial ground. *See* Ground, radial
Radio frequency interference (RFI), 70, 122
Raw power (term to avoid), 35
Raw utility power (term to avoid), 35
Reactance, 63-64
Receptacle circuit testers, 32, 109
Receptacle miswiring, 152
Recovery time, 32
Recovery voltage, 32
Rectifiers, 40, 61, 71, 89, 176, 178, 189-190
Redundant systems, 181-182
Relative humidity measurements, 122, 140
Reliability, 68, 181, 183-185, 187, 192
Resistance, grounding electrode, 128-129
Resistance, lumped, 57
Resistance testers, 104-105, 110-111
Resonance considerations, 64-68, 76
  conductor self-resonance, 66-68
  parallel resonance, 65-66
  series resonance, 64
Restoration, system, after failure, 196-197
Reverse transfer
  automatic (UPS), 189
  impedance, 31

RFI/EMI (radio frequency interference/electromagnetic interference), 44, 49, 51-52, 70, 122, 140, 205
  case history, 155
Ring-ground, buried, 206, 229-230
Ring Wave, 60, 92
Ripple current, 180
Root-mean square (RMS) ammeters, 107
Root-mean square (RMS) voltmeters, 106-107
Rotary UPS, 174-176
Rotating transformers, 171-172
RS-232 port, 138

S

Safety considerations, 22, 46, 50, 199-200, 205
  case history, 156
  grounding, 22-23, 90, 205. *See also* Grounding
  listing avoidance, 192
  protective equipment, 103, 108, 116, 127
Safety ground. *See* Equipment grounding conductor
Sags, 70, 105, 162, 171, 241-246
  case history, 143-144
  definition of, 32
  protection, 49
Sampling devices, 107
SDS (separately derived AC system), 36
Semiconductors, 89, 150
Sensitive electronic equipment
  case histories. *See* Case histories
  computers. *See* Computers
  harmonics. *See* Harmonics
  introduction and background, 21-24, 39-46
  performance, 200, 205
  power source interactions. *See* Load and power source interactions
  shielding. *See* Shielding
  standards, 21, 51-54. *See also* IEEE
  see also Equipment specification and selection
Separately derived AC system (SDS), 36
Series resonance, 64
Shared circuit (term to avoid), 35
Shared ground (term to avoid), 35
Shielding, 204, 226
  definition of, 33
  electromagnetic, 95
  electrostatic, 94, 161
  EMI/EMC areas, 51-52
Shunt capacitors, 59
Signal diodes, 89
Signal lines. *See* Data lines
Signal reference grid (SRG), 210, 212-216
Signal reference plane (SRP), 222
Signal reference structure (SRS), 27
  definition of, 33
  grounding, 91-93, 210, 212, 216
Single-phase systems and loads, 200, 204

Single-point grounding, 214-215
Site surveys and site power analyses
    applying data, 139-141
    approaches, 45-46, 123-124, 142
    conducting, 125-141
    coordinating involved parties, 124-125
    distribution and grounding
        system, 124-135
    documentation, 130-135
    environment, 140-141
    instrumentation, 136-139. *See also*
        Instrumentation
    levels 1 to 3, 125-126
    long-term power monitoring, 141-142
    objectives, 123-124, 142
    voltage quality, 136-140
Skin effect, 74, 77, 96
Slew rate, 33, 72
Smoke/fire detector system case history, 156
Soil resistivity, 82, 86
Solidly grounded systems, 205, 207
Solid state devices, 40, 49, 146, 148, 176-180, 205
Specifying engineer's considerations,
    187-189
Spectrum analyzers, 104-105, 120
Spike (term to avoid), 35
Square-law type voltmeters, 107
SRG (signal reference grid), 210, 212-216
SRP (signal reference plane), 222
SRS. *See* Signal reference structure
Standards, 21, 51-54. *See also* IEEE
Standby power systems, 161-163, 172-174
Star-connected circuit, 33
Star ground. *See* Ground, radial
Start-up current values, 109
Static transfer switch, 184
Static UPS, 176-177
Steady state voltage distortion, 73-77
    potential impacts of, 74-77
    sources and characteristics, 73
Steel mill casting shutdown, case
    history, 146-147
Step loads, 71
Stray reactive coupling, 84
Structural steel, 205-206, 213, 215, 217
Subcycle outage (term to avoid), 35
"Super isolation" transformers, 163
Surge. *See* Transient voltage disturbances
Surge reference equalizers, 33, 111-115
Surge suppressors, 43, 160, 162, 165-166, 205
Surveys. *See* Site surveys and site power
    analyses
Swells, 70, 105, 162, 171, 241-244
    definition of, 33
Switchboards, 229
Switching, transfer, 205
Switching surges, 22, 78-80, 205
Synchronization test, 194
Synthesizers, magnetic, 169-170, 173
System configuration, 181-182
    isolated redundant systems, 182-183
    parallel systems, 181-182
System load rating, 178

T

"Tank" (LC) circuits, 65
Tap changers, 166-167
Technical Committees (IEC), 53-54
Telecommunication system lightning
    protection, 221-225
Telephone network, 50, 214, 216-217
Temperature measurements, 122, 140, 156-157
TEMPEST requirements, 51-52
Testing, verification. *See* Verification
    testing
Text monitors, 114-117
Thermocouple type voltmeters, 106-107
Three-phase systems and loads, 200, 204
Thunderstorms, 145-146. *See also* Lightning
    protection; nonarcing atmospheric
    charge redistribution
Thyristors, 153-154, 167, 189-190
Tolerances, electronic load, 70
Topological grounding methods, 52
Total harmonic distortion, 26, 188
Transducers, 112
Transfer characteristics, 189
Transfer switching, 22, 78-81, 205
Transfer test, 194
Transfer time (UPS), 34, 189
Transformers, 58-59, 61, 71-72, 106, 176-177
    ALVRTs, 231
    constant voltage, 168-169
    derating, 74-75, 232
    dry type, 231-236
    eddy current heating, 74-76
    ferroresonant, 168-169, 173, 177
    impedance, 58
    isolation, 154, 160-164, 172, 207-208
    K-factor rated, 75-76, 233
    rotating, 171-172
    sizing, 127
    "ultra isolation," 163
Transient voltage disturbances, 22, 49, 59-60,
    77-89, 105, 171
    buried cables, interaction with, 86-87
    case histories and power quality studies,
        146-149, 241-244
    common mode, 162
    coupling mechanisms, 84-86
    definition of, 34
    electrical fast transient (EFT), 61, 89
    frequency distribution, 88-89
    lightning. *See* Lightning protection
    normal mode, 162
    overhead conductors, interaction with, 87
    potential impact, 87-88
    power system switching, 22, 78-81, 205
    sources and characteristics, 70-72, 78-84
    see also Notches; Overvoltage; Swells
Transient voltage surve suppression devices.
    See TVSS devices
Transistors, 89
Transverse-mode noise, 34, 162
Triplen harmonics, 73, 76, 150, 230
True RMS voltmeters, 106-107

INDEX

TVSS (transient voltage surge suppression) devices, 49, 84, 94, 148, 150, 154-155, 207, 218-223, 225
Type I, II, III power disturbances (terms to avoid), 35

U

"Ultra isolation" transformers, 163
UL (Underwriters Laboratories), 52-53, 75, 192, 201-202
Unbalance, 42
Unbalanced load regulation, 34, 188
Underground distribution systems, 86-87, 245-246
Undervoltage, 34, 70, 105, 162, 245
Underwriters Laboratories (UL), 52-53, 75, 192, 201-202
Uninsulated grounding conductors, 218
Uninterruptible power supply (UPS), 161-162, 181-185, 187-189, 193-195
  batteries for, 179-180
  case histories, 146, 149-156
  grounding, 217
  inverter technology, 174, 190
  lightning protection, 220
  rectifier technology, 189-190
  rotary, 174-176
  static, 176-177
UPS. See Uninterruptible power supply
Upset effects, 48
Utilities, 22, 39, 41, 43, 68
  capacitor bank failure case histories, 151-152
  capacitor banks, switched, 139-140, 146-147
  fault clearing, 145
  site surveys, involvement in, 125

V

Varistors, 154, 160, 165
Vendor-supplied specifications, 191-193
Verification testing, 193-195
  AC input failure and return test, 194
  efficiency test, 194
  harmonic component test, 195
  load tests, 194-195
  overload capability test, 195
  synchronization test, 194
  transfer test, 194

visual inspection, 193-194
Voltage, recovery, 32
Voltage distortion, 72, 136, 188, 202-204
  definition of, 34
  rate of change in, 48
  steady state, 73-77
  types of, 162
  see also Frequency deviations; Interruptions; Overvoltage; Sags; Swells; Undervoltage
Voltage range, input, 28, 188
Voltage regulation, 23, 72, 188
  capacitors, 147, 151
  definition of, 34
Voltage regulators, 160, 162
  buck boost, 167-168
  constant voltage transformers, 168-169
  tap changers, 166-167
Voltage (supply system), selection of, 200-201
Voltage surges. *See* Transient voltage disturbances
Voltmeters, 104-107
  root-mean square (RMS), 106
  true RMS, 106-107

W

Watthour meter, 42
Waveform analyzers, 117-120
Waveform variations, 42, 201
Wear, component, 195-196
Wet cell batteries, 179, 195
Wiring, 209, 230
  computer rooms, 201-202
  costs, 186, 226, 228
  devices, 236-237
  losses, 226, 228
  measurement instruments, 103
  problems, case histories, 151-154
Working Groups (IEEE), 24, 45
Wye connection, 33

Y

Y connection, 33

Z

Zeners, 89